THE DIALECTICAL UNIVERSE

SOME REFLECTIONS ON COSMOLOGY

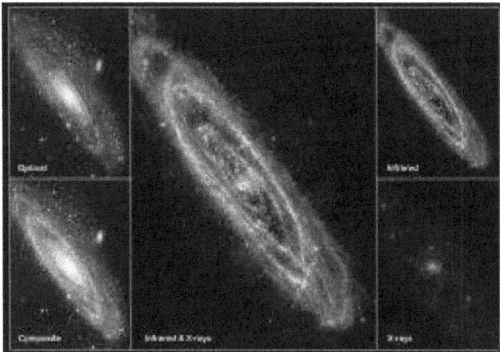

ABDUL MALEK, Ph.D.

THE DIALECTICAL UNIVERSE

SOME REFLECTIONS ON COSMOLOGY

ABDUL MALEK, Ph.D.

Agamee Prakashni

Published By: Agamee Prakashani

First Published: February 2012

Printed by: Swarbarna Printers

Price: US $15.00

ISBN No.: 978 984 04 1444 4

Cover: Andromeda Galaxy: The middle picture is a composite image in IR and X-ray, the other pictures clockwise from top left are optical, I.R., X-ray, and a composite image of IR, optical and X-ray Credits: infrared: ESA/Herschel/PACS/SPIRE/J. Fritz, U. Gent; X-ray: ESA/XMM-Newton/EPIC/W. Pietsch, MPE; optical: R. Gendler

This Booklet is dedicated to the memory of Prof. Satyendra Nath Bose, one of the greatest scientists of 20th century. He ranks among the few personalities in history, who wrested recognition for eternity, against all odds, by the sheer prowess of their intellect and their iron will.

Prof. Bose made profound contribution in natural sciences, particularly in quantum statistics and mechanics, and also in socio-politico-cultural fields while being a member of the faculty of the University of Dhaka. His self-confidence and tireless determination to win recognition for his discoveries from a cold, capitalism dominated, chauvinistic and Euro-centric natural science will provide courage to any aspiring person even in the most obscure corner on this planet.

This author was highly inspired, both as a student and later as a member of the faculty of the University of Dhaka by the examples left behind by Prof. Bose. This Booklet - a result of decade-long private and amateur work by this author in the field of astrophysics and cosmology, is a tribute to the immortal legacy of Prof. Bose.

CONTENTS

FOREWORD

The dialectical mode of thought evolved early on in the evolution of human civilization in (dialectical) unity and opposition with causality, formal (Aristotelian) Logic and monistic reductionism which G.W.F. Hegel (1770 – 1831 A.D.) collectively called the "view of understanding".

A crude sense of short-term cause and effect is common in all life forms, including the most primitive ones. This quality roughly separates life from non-living matter. Simple movement & mechanics even in primitive life forms is impossible without a general sense of causality. With the progress of evolution, this sense of causality is accentuated culminating in its highest development in man. The "view of understanding" represent the "good old common sense" acquired through everyday life experience and passed on and attenuated through generations that eventually attains a generalized "instinctive" form of judgment common to most members of a society or an epoch The dialectical mode of thought on the contrary, could only arise with the higher capability for communication, abstraction, introspection, and reflection etc. that could manifest itself in the highest developed form of matter - the thinking brain of man.

The "view of understanding" or (more commonly known) formal logic in the western world has its roots in early Middle Eastern and Greek philosophy, particularly those of Plato (429 – 347 B.C.) and Aristotle (384 – 322 B.C.) and developed mostly in Europe through early-Greek idealism, monotheistic theologies, medieval scholasticism, the rationalism of the Enlightenment, Empiricism and classical materialism culminating in Emanuel Kant (1724 – 1804 A.D.). The sharp class conflicts and the development of natural science in Europe shaped the formation of this view. After the bourgeois democratic revolution, it became the most dominating narrative not only in Europe, but also around the world, as European colonial and economic domination intensified.

The most characteristic feature of the "view of understanding" is that it denies the pervasiveness of contradiction in nature, history or thought. For it everything is created "perfect in itself", that only has extension in space, but no change in time. Aristotelian doctrine of Identity, opposition and the excluded middle is the cannon of its procedure. For crass materialism specially for official natural science there is no freedom of will, everything from the creation of the universe according to the anthropic principle of the Big Bang theory to biological determinism through genes are predetermined and follow their course in time. It admits only static dichotomy between the unchanging opposites, but no unity or reversibility between them. Motion (change, development, evolution, transformation etc.) in this view can only be caused by an omnipotent external agency - the "unmoved mover". There were of course elements of dialectics in this narrative at least since Heraclitus (544 – 483 B.C.) and other early Greek philosophers like Aristotle (384 – 322 B.C.) through to Epicurous (341 – 270 B.C.) and their followers, but those were not emphasized or at best not recognized. Idealism, theology, rationalism, classical materialism, British empiricism and Newtonian mechanics are the highest representations of this worldview.

The aim of this world outlook is to know au unchanging,, eternal, certain, absolute, causality based pre-existing world - the so-called objective reality that is independent of man. The task of this epistemology of both idealistic and materialistic varieties (natural science included) was to find that absolute reality. For idealism, that reality is the thought world, which has no independent existence, but can only be perceived by a rational mind; the material world is only a temporary chimera, or "Maya". For materialism and official natural science on the other hand, the material world is the only reality that can be understood by empirical and scientific investigation.

2

The dialectical way for an epistemic understanding of the world arose in all major early civilizations, Middle Eastern, Greek, Indian, and Oriental etc. Because of their emphasis on spiritualism, most Eastern dialectical modes of thought more or less took the form of absolute dualism, monism, a cyclical or of a static nature, Although the basic attributes of dialectics i.e., unity of the opposites, change, development etc., were recognized, they became confined within a bigger unity and cyclical forms – the Yin-Yang of the Orient or as in the case of Buddhism, repeated and progressive cycles eventually leads to the unchanging and absolute unitary state of Nirvana or as in Hinduism, the creation (Brahma), destruction (Shiva) and preservation (Vishnu) are the three eternal and simultaneous attributes of the same unitary deity.

Only one form of ancient dialectics, the helical or spiral form enunciated by the great Greek thinker Heraclitus of Ephesus (544 – 483 B.C.) has borne real fruits. His assertion that everything is in constant change, as a result of inner strife, posited the brilliant germ for the subsequent development of dialectics. G.W.F. Hegel (1770 – 1831A.D.) was the first to present its working in a comprehensive and conscious way as the general laws of motion & development of nature, history, society and thought. Karl Marx (1818 – 1883 A.D.) and Frederick Engels (1820 – 1895 A.D.) retooled Hegelian dialectics into its materialist form and used it in all branches of knowledge including natural science, as history's most powerful and far reaching epistemological vision for an understanding of the world and for systematically using that knowledge to benefit humanity and life in general.

Hegel's dialectics is the outcome of the unprecedented development in the economic, social, philosophical and natural sciences triggered by the Copernican revolution in cosmology in 1543 A.D. and the ensuing great bourgeois democratic revolution in Europe in general. The paradigm of the "view of understanding" that was containing these revolutionary developments was tearing at its seams. The absoluteness, perfect ness, rigid categories, beauty, goodness, non-contradiction etc. of the "view of understanding" began to dissolve in the fluid of new developments. Ironically, the two most important Elements that the "view of understanding" banished from its vocabulary, namely, the idea of *evolution* and *contradictions,* were to undo it and engender Hegelian dialectics.

The idea of evolution was gaining strength by early 18th century. J.B. Lamarck (1744 – 1829 A.D.) for the first time, proposed the idea that evolution in the animal world proceeded according to natural laws; and his contemporary philosopher Immanuel Kant (1724 – 1804) proposed the "nebular hypothesis" for the evolution of the cosmic bodies. Also, the "view of understanding" was rife with contradiction and conflict from the start, first between theological scholasticism and rationalism and later between rationalism and the new radical currents of empiricism (both materialist and idealist), materialism and natural science. Due to the practical necessity of life and society, Judeo-Christian theology had to introduce contradictions in its narrative, such as evil, "freedom of the will", miracles etc., in Gods perfect world that were repugnant to rationalism. Rationalism and classical materialism (including natural science) which also shunned contradictions and used causality as a tool for enquiry had no answer to these and other contradictions in the material world, either. The same situation persists even today in official science – natural, social and philosophical.

The issue came to a head by the middle of eighteenth century when David Hume (1711 – 1774 A.D.) questioned the certainty of human knowledge and the capability of human mind to discern reality. He also attempted to show that some of the most fundamental conceptions of contemporary epistemology (of the view of understanding) such as causality, identity, substance etc. are spurious, because they lack *necessity* and *universality.* Causality also leads to a *first cause* (i.e., the effect of a cause that is unknown) and hence a mystery, such as the God of theology or the Big Bang theory of modern official natural science.

4

Immanuel Kant (1724 – 1804), who wanted to refute Hume, eventually was forced to agree that formal logic is incapable of finding the truth about the noumena or *things-in-themselves*. Formal logic he argued can only know the phenomena – the appearance that we perceive through the senses. In developing the antinomies of reason, Kant showed that formal logic always leads to contradictory conclusions and contradictions. Hence, Kant posited, that reason can never know reality, and therefore, in order to avoid contradictions our reason must limit itself to structuring and manipulating its subjective intuitions: its logical categories, and two types of a priori knowledge –analytic a priori and synthetic a priori. , Kant warned philosophy to abate its claims; it must give up all attempts to know reality, to penetrate behind the veil of appearances.

It fell on Hegel to restore the honor of philosophy as the science of all sciences, the soul of all knowledge, by pulling it out of its intractable problems and the lowest moment it reached by the time of Kant. And strangest of all, this he accomplished by embracing the very same elements, namely the ideas of evolution and contradictions etc., which the "view of understanding" abhorred the most. On the contrary, he put them at the very heart of his new philosophical system – the dialectical method.

Hegel unambiguously rejected the law of non-contradiction of theology, old idealism, rationalism and classical materialism, the "excluded middle" of Aristotle and the *thing-in-itself* of Kant. For Hegel absolutely everything depends on *"the* identity of *identity and non-identity."* Opposites reside together in the very element of a thing or a process in simultaneous unity and opposition to each other and a resolution of this logical contradiction and conflict provides the dynamics for change, motion, evolution, development etc. Modern physics so far have failed to find a unitary process or fragment of matter (such as so called "magnetic monopole") without its dialectical opposite even at the sub-atomic and sub-nuclear levels. Particles at present considered being unitary (fundamental) such as, mesons, quarks etc. turn out to be composite of opposites. The purest and

5

the holiest form of material existence i.e., light (photons) like mesons is now seen to be composites of the dialectical unity of the opposites of matter and antimatter particles residing in its unit and can be resolved into them under suitable circumstances. A photon of proper minimum energy (gamma ray) under certain condition will be converted to a pair of electron and its opposite antiparticle, positron.

For Hegel, there is no absolute *being* or absolute *nothing* by themselves, they always contain each other and so one can issue out of the other without breaking the rules of formal logic. *"As yet, there is nothing and there is to become something. The beginning is not pure nothing, but a nothing from which something is to proceed; therefore being, too, is already contained in the beginning. The beginning, therefore, contains both, being and nothing, is the unity of being and nothing; or is non-being which is at the same time being, and being which is at the same time non-being."* Hegel thus posited his dialectics at the very core of reality and ontology itself. Everything in the universe, therefore, implicitly contains everything else and hence can be derived by evolution from each other, without the necessity of any act of creation!!. Hegel performed this miracle (which the "view of understanding" was unable to do without resorting to the intervention from Providence or God), ironically, without the breach of the fallacy of illicit process or the principle: *ex nihilo nihil fit* of formal logic, that forbids the derivation of a conclusion which is not present in the premise or a consequent not contained in the antecedent.

Modern quantum electrodynamics and spectroscopy proves Hegel's dialectics experimentally with the highest possible degree of precision. For quantum mechanics absolute vacuum (nothing) is impossible – vacuum is teeming with ghostly "virtual" particles "constantly coming into being and passing out of existence" This phenomena causes observable effects, for example in the spectra of atoms as the "Lamb Shift" which can be measured with a precision of one part in a million. Some of these "virtual" particles can become "real" through some

6

quantum processes such as "tunneling" effect etc. or if enough energy is available. This and other quantum mechanical phenomena such as wave-particle duality are inconceivable for the view of "understanding".

One of the most important characteristic of dialectics is that it denies any permanence or absoluteness in any thing or a process, everything is in a flux of coming into being and passing out of existence so that *change* (with infinite temporary stages) remains the only absolute. For it, contradiction (unity of the opposites) in the unit of a thing or a process is the most fundamental attribute of all existence (of material or thought objects) and *change* or motion is the manifestation of that inherent contradiction. While the "view of understanding" is cognitive, dialectics is innovative. Unlike the "view of understanding", the aim of dialectics is not to know a fixed and pre-existing reality or absolute truth, but to reflect the ever-new realities ushered in through the inner conflict within the old and mediated by chance and necessity as the dialectical process of the world moves on in stages, without ever terminating in a final point.

Historically, with the dissolution of the primitive communistic societies into (economic) class-based ones, along with their hierarchical socio-political superstructure, the "view of understanding" as opposed to dialectics necessarily, gained predominance in social dynamics as well as in epistemology. Because of its very nature, as the conservative, the resisting, the preserving side of what exists, the "view of understanding" always was the natural choice for class based status quo and the established order of the time, while dialectics represented the revolutionary side; because dialectics denies the stability or the permanence of what exists. This conflict continues till today and has only intensified especially after the First World War, when the Anglo-American led monopoly capitalism evolved as the most dominant politico-economic force on the world stage.

In natural science, the "view of understanding" prevailed exclusively without any opposition until Hegel and till now

7

continues to be the only form of discourse for official science, governed by world monopoly capitalism. The "view of understanding" was adequate for classical and Newtonian mechanics and old materialism. Hegel for the first time engendered conscious dialectical approach not only to philosophy & history but also to natural science. The present knowledge on the evolution of heavenly bodies such as galaxies, of terrestrial nature, evolution of life, quantum mechanics, the discrete, discontinuous & quantized structure of matter from the subatomic to the cosmic etc. can now be rationally understood only from the dialectical point of view. Engels could put relatively more efforts on the elaboration of dialectical approach to an understanding of natural science and helped clear up much of the confusion that was rampant among the natural scientists of his time including the newly enunciated theory of evolution by Charles Darwin (1809 – 1882). But after Marx's death in 1883, Engels could only sporadically keep up with the developments in natural science, because of other more urgent tasks at his hand. His two immortal works: Anti-Dühring and Dialectics of Nature (which he left as an unfinished manuscript) remain the invaluable treasures and a beacon of light for a dialectical approach to acquire positive knowledge of nature.

By late 19th Century and immediately following the great discoveries of Marx and Engels in socio-politico-philosophical, historical and economic sciences and those of Darwin in biological sciences, revolutionary developments in biology, paleontology, geology, chemistry, and particularly physics (radioactivity, X-ray, quantum phenomena, spectroscopy, atomic theory etc) vindicated dialectical materialism and put epistemic view of "understanding" of classical materialism, Newtonian mechanics and theology into a tailspin.

The revolutionary dialectical materialism of Marx and Engels not only elicited violent reaction from the incipient monopoly capitalism in Europe but it also induced a regressive and a virulent counter-revolution in official natural science, as regards to its approach to materialism, classical (Newtonian) mechanics and to British empiricism in general. Classical materialism gave rise to sterile determinism and reductionism, anti-metaphysical Newtonian mechanics was replaced by Einsteinian mathematical idealism and (materialistic) Lockean empiricism degenerated to idealistic Neo-Berkeleyan and Neo-Kantian positivism of all kinds. Official natural science has remained hostile to dialectical materialism ever since.

Monopoly capitalism met the crisis in its ideological base by repudiating materialism and by co-opting idealism, theological fundamentalism and positivism. It found in Einstein's mathematical idealism especially his theory of General Relativity (GR) a convenient ideological crutch to prop up its fledging superstructure and theology. Neo-Berkelean positivism of the Vienna Group in philosophy and GR in natural science were twins born as if " made to specification" for monopoly capitalism; the same way as Christianity for the Roman Empire evolved out of the philosophy of Plato and the Jewish mythology of that period. Thanks to the collapse of the Bolshevik revolution in the Soviet Union and a near monopoly in the control and practice of natural science, official science retains its tenuous hold by using the mystical authority of GR and still successfully fends off the role of dialectics; even though the more recent developments in natural sciences collectively continue to gravely undermine the logical foundation of mathematical idealism.

 After Engels' death, V.I. Lenin (1870 – 1924 (A.D.) continued the works of Engels in combating the virulent surge of idealism in natural science particularly personified by Ernst Mach; at the advent of the revolutionary discovery of quantum phenomena in physics. Lenin demolished the idealist and reactionary philosophy of Ernst Mach et. al, and of positivism in general, with regard to natural science and the quantum phenomena, in his seminal work- "Materialism and Empirio-Criticism". But like Engels, he could do it only as a secondary effort, because the urgent task of leading the Bolshevik revolution and establishing the first socialist state totally preoccupied him. With the defeat of the Bolshevik revolution, the powerful voice of dialectical materialism in natural science was silenced. The "view of understanding" as represented by classical materialism in combination with Einstein's mathematical idealism remains the dominant ideological basis of official natural science.

But in spite of intense hostility from official science towards the dialectical approach, scientists like J.B.S. Haldane, J.D.Bernal, and some Soviet scientists, particularly I.Oparin continued the great tradition set forth by Engels in the field of biology. More recently, Stephen Jay Gould, Richard Lewontin, Richard Levin, Steven Rose et al (in the English speaking world) and others have used dialectics as a powerful tool in the field of biology, even though they are generally ignored by the mainstream. Only in the field of physical sciences the domination of the "view of understanding" remains unchallenged. Although the development of quantum mechanics delivered a mortal blow to the philosophical base of classical materialism and causality, official natural science opportunistically uses the colossal technology spurred by quantum mechanics, but loudly denies its reality and dismisses it as "weird". The theory of evolution and quantum mechanics are now so well established that official science no longer tries to suppress them. But instead uses its dominant authority to render them harmless, impotent and blunted, by shepherding them into simplistic, mechanistic and reductionist generalizations - "absolute truth", "theory of everything" etc ad nauseum.

Because of its total monopoly on the funding and research in the field of astrophysics & cosmology and its medieval-like control and intolerance of dissident views, the hold of monopoly capitalism on natural science remain as formidable as ever. After the phenomenal development since the Copernican revolution, natural science comes back to the correspondingly higher helical position it had during the medieval period of history. Anthropomorphic and geometry based cosmology and absolutism once again becomes the focal point and the source of the power of despotism and class antagonism on an international scale. Einstein's theory of General Relativity (and the derived

10

Big Bang theory) stands in the same relation to modern time as Ptolemy's (323 – 283 B.C.) Epicycles and geocentric cosmology was to the medieval despotism. Once again, divine authority and absolute truth of the universe is claimed in the name of a geometry based theory whose successes consist of its ability to retrospectively fit observations by using adjustable parameters and by invoking mystical objects & processes. Evidently, classical materialism has driven itself into a blind alley and exhausted all its revolutionary potential to propel natural science any further ahead. It is time for dialectical materialism to take that mantle and usher in another great leap forward for humanity.

This monograph is a collection of articles and views on astrophysics and cosmology, some of which were published earlier in Internet based journals and websites. Each article is a stand-alone type on a particular topic and can be read independently of the others. Unfortunately, this format necessarily can cause some overlap and repetition of ideas. Based on a dialectical and quantum mechanical point of view this monograph is intended as a critique and exposé of the current state of affairs in the field of cosmology and astrophysics, and also as a potential challenge to the fundamental philosophical basis of modern official natural science in general. Any serious challenge must provide its own alternatives to be credible. It is therefore, the task of dialectical materialism and its adherents to offer such alternatives. Some humble and tentative alternative proposals on some aspect of cosmology and astrophysics are offered in this monograph, with the hope that more profound dialectical insights in the future will push these to obscurity.

Montreal, September, 2011 **Abdul Malek**

11

PROLOGUE

The universe is the dialectical manifestation a) of matter (the *Substance* –i.e., "the sum total of Extension and Thought" of Spinoza) in eternal self-propelled motion, b) of its infinite series of leaps, change, transformations and historical processes of development; c) of the perpetual process *of its objects coming into being and passing out of existence* mediated by blind chance and an iron necessity that is inherent in chance. This motion is reflected in three perceived general laws i.e., i) inter-transformation of quality and quantity; ii) interpenetration of the opposites; and iii) the law of the negation of the negation that follow the dynamical and helical triads of thesis, anti-thesis and synthesis. Matter itself appears and disappears in the void of the infinite universe, in some elementary forms as dialectical and quantum mechanical necessities and in accordance with the first dialectical triad of *being-nothing-becoming*. The reason and the mechanism of this motion lie in the internal and inherent logical contradiction (dialectics) and tension auto-developing within matter itself and not due to an *impulse* from without. Outside factors can however, in a limited way influence the course of the development of a particular and isolated element or process under specific conditions within narrow space-time limit in the sense of cause and effect, but in all generality chance and necessity hold their sway. Life, humanity and thought likewise evolve (as part of the universal motion of matter), through an inter play of chance and necessity and through infinite discrete dialectical leaps of historical processes.

An epistemological understanding of this eternal motion – i.e., its reflection in human mind also evolved as similar dialectical and historical processes. After about 4000 years of recorded history of human practical activities, thought and social development and especially after the great leap forward of the bourgeois democratic revolution in Europe since Copernicus (1473-1543); that led to an unprecedented scientific and social development; concrete objective conditions were created for a great mind to perceive, comprehend and to formulate in an all-round way the general laws that seem to underpin the eternal motion of matter. That mind was provided in the person of G.W.F. Hegel (1770-1831 A.D.).

Hegel distilled out the essence of the last few thousand years development of history, society and of social/natural philosophy and brought into sharp and concrete focus the profound insights that were dimly perceived by the previous philosophers, particularly the great ancient Greek thinker Heraclitus (544-483 B.C.). But what to idealist Hegel is the movement of his "Absolute Idea" as a distinct entity alienating itself in material Nature and coming back to itself again through the consciousness of man and through giant leaps (in an apparent suspension of unity of matter and thought) in a closed loop formation; for Karl Marx (1818-1883 A.D.) consciousness/Idea/thought is the manifestation of the very same matter itself in its perpetual motion. Thus for Marx like Spinoza (1632-1677 A.D) (and unlike Hegel) matter and mind/consciousness are but two different attributes of the same entity i.e., *Substance* and not (as for Hegel) two distinct phases i.e., the "Absolute Idea" separated and towering over material Nature. Marx thus put Hegel's philosophy on a materialist and a firmer dialectical foundation and eliminated the dialectical inconsistency of abject matter separated from the supreme "Absolute Idea" as apparent in Hegel. Every manifestation of one is but a manifestation of the other, seen under different aspect.

The contribution Marx and Engels made to philosophy (as it is aptly called dialectical materialism) is that, it bridges the gap between classical materialism and idealism, extracts the positive elements from both and unveils the dialectical unity and conflict of these two opposites that operate in the actual development in the history of human society and philosophy. It also exposes more clearly, the limitation and hollowness of British empiricism, both in its idealist (Berkeley) and materialist (Locke) forms as both are based on the dichotomy of the polar

13

categories of "Identity, Opposition and the Excluded Middle" and the "view of *understanding*" of formal logic. Crass materialism (which is more prevalent in natural science) sees matter as being all-important and mind a passive blank board (Tableau Blanc) on which impressions or images is made by matter. The only role of mind is to work up these images to form concepts, ideas etc. The epistemology of classical materialism based on the *understanding* of "good old commonsense" of everyday experience is that, an "objective reality" of absolutely fixed and exactly quantifiable, definable character exists and it is independent of the mind. While idealism on the contrary asserts that matter as such does not exist, the "physical reality" which materialism considers as primary is but an illusion; it is the mind that is the primary active agent. The mind imposes characteristics and properties on various things in the phenomenal world. But what both materialism and idealism miss in their discourse is the most fundamental question of any worthy epistemology, i.e., the question of "motion". For both, any motion, change, development is due to an "impulse" from without, from an omnipotent "unmoved mover" or God.

Immanuel Kant (1724–1804 A.D.) took the opportunistic middle ground to bring an unprincipled reconciliation between the two antagonistic opposites. Like materialists he apparently admitted the existence of an objective reality (thing-in-itself) that is independent of the mind, but declared that it was unknowable. While like the idealists he asserted that the mind is the active agent that projects (a priory) space, time, causality, pure reason etc. within which the phenomenal world operates. But to the greatest merit of this giant thinker and unlike his pigmy followers of empiricism, positivism, and either/or dichotomy of classical Idealism/Materialism; Kant for the first time advanced the dialectical perspective of the universe through his "nebular hypothesis" and discovered the categories and the laws that operate in the development of "Thought".

Dialectical materialism asserts that an "objective reality" independent of consciousness does exist and the mind is a part of it. And that classical materialism has a limited, short term, one-sided, dogmatic, naive and a very narrow perspective of the "objective reality". For dialectical materialism (contrary to classical materialism) this reality is always in flux, motion, change, development and it is unstable and inherently uncertain at quantum scale, such that an exact, fixed, definitive and quantifiable understanding or an exhaustive description of this reality forever is impossible. But this also is at the root of the basic dialectical contradiction and the unity of the opposites between ontology and epistemology, and resolves this contradiction in the endless progressive evolution of consciousness/mind. This process can never stop or come to an end, terminating to some absolute truth. Idealist Hegel in violation of his dialectical laws contrived to resolve/eliminate this contradiction in his "Absolute Idea". The results accumulated so far by natural, social and philosophical sciences in the realm of the quantum, micro, cosmic and terrestrial nature; life; society; and thought, are in conformity with the dialectical materialist view of the world.

Official natural science still pursues the arbitrary, random and the metaphysical approach in its investigation of Nature. For every inter-connection, transition, evolution etc. in Nature and society, fantastic fabrications created in the mind of scientists or philosophers are substituted for the real ones. But the empirical evidence collected by natural science itself in the realm of the terrestrial Nature provides incontrovertible proof of the dialectical evolution of the earth from its early planetary form through geological, oceanic and atmospheric formations up to the present time. Laboratory experiment simulating the early atmospheric conditions of the young earth provides evidence for the formation of organic molecules that were essential for the later evolution of life and the formation of the cell. I.Oparin and J.B.S Haldane gave convincing insight as to how the oceans provided the media where the organic molecules formed aggregates that eventually led to primitive life forms through dialectical chance and necessity. It is Darwin's theory of evolution and modern paleontology that vividly shows the dialectical interplay of heredity and adaptation, chance and necessity etc. in the origin and the evolution of the species. Quantum mechanics, which dramatically display the role of chance and necessity, gave the most devastating blow to causality of metaphysics. The concept of a creator/regulator in the realm of terrestrial Nature and in the micro-world is now abolished forever.

And in the realm of society and history, Karl Marx showed how man created himself and history through his labor, and how the dialectical interplay and the class struggle of the productive forces and the relation of production led to the evolution of society, history and thought and how the "surplus value" of unpaid labor created the edifice of capitalist wealth and dispossessed the proletariat. Engels pointed out the importance of the role played by labor in the transition of ape to man; and how the dialectical and simultaneous interdependent development of the hand, speech and the brain evolved through labor. Based on the materialistic conception of history of Marx and independent Anthropological discoveries of Lewis H. Morgan; Engels also elaborated the dialectical perspective on the origin of the Family, Private Property and the State and showed how production and reproduction of immediate life constituted the determining factor in the history and the immediate pre-history of man.

An understanding of the major links in the dialectical evolution of the terrestrial Nature, life, history and thought has now been completed in their main general outline such that it allows no room for a "first cause" or a creator of all things any more in this realm. This evolution is the result of innumerable dialectical and qualitative leaps engendered by chance and necessity. The role of modern scientific and historical (both natural, paleontological and social) research in the realm of terrestrial nature is to fill up the details in this general picture in an ongoing and never-ending effort and to use the tools of dialectics to extend and acquire positive knowledge in the realm of the macrocosm and microcosm.

Marx pointed out the dialectical unity of the opposites of matter and mind – the fact that mind/consciousness is a property of matter itself. Therefore the subjectivity of the mind is no less valid than the objectivity of the world, because they are the parts of the same unity of the opposites. The subjective aspect of Nature comes into play early on in the evolution of primitive life forms (an amoeba for example will react spontaneously to stimulus in a way to try to preserve itself; a plant is an active agent in creating the conditions of its life and growth) and progresses dialectically & historically with the evolution of higher life forms. With the evolution of man the subjective aspect becomes increasingly more and more forceful in the realm of socio-economic and historical development. The "subjective will" of man has validity as a material force like other objective material forces of Nature.

The abstract idea of the "freedom of the will of the Spirit" of Hegel and of Idealism gets a totally different meaning in the dialectical materialism of Marx. As Engels wrote (Anti-Dühring, p-125), "Freedom does not consist in the dream of independence of natural laws, but in the knowledge of these laws, and in the possibility this gives of systematically making them work towards definite ends. This holds good in relation both to the laws of external Nature and to those which govern the bodily and mental existence of men themselves – two classes of laws which we can separate from each other at most only in thought but not in reality. Freedom of the will therefore means nothing but the capacity to make decision with real knowledge of the subject. … Freedom therefore consists in the control over ourselves and over external nature, which is founded on knowledge of natural *necessity*; it is therefore necessarily a product of historical development. The first men who separated themselves from the animal kingdom were in all essentials as unfree as the animals themselves, but each step forward in civilization was a step towards freedom".

To their credit, the idealists (even though inconsistent with their narrative) allow some role of subjective "free will" on ethical issues and allows some minor contradiction in their epistemology by invoking the role of "evil" or "Satan" as opposed to the "good" of God. For hard-core "materialist" ideologues "free will" is nonsense; everything is pre-determined according to the "anthropic" principle, as their mathematical idealism "proves" it to them and follows the course dictated by the all-powerful conditions of the so-called "objective reality"; the only subjective thing for them is to opportunistically use these conditions for "enlightened self-interest". For them, there is no role at all for chance and necessity; there is no contradiction in the world, only a sharply divided and a clear-cut choice of yes/no, good/bad, and positive/negative etc. dichotomies. But when they dig too far into the "objective reality" with their anti-dialectical and faulty tools of causality, they discover with horror that at the root of their static "objective reality" there is nothing but the omnipotent hand of the theological God!

This pathetic drama is most visible in the position that official natural science takes, for example, on the questions on biology, evolution, quantum phenomena, the origin and the evolution of galaxies and (the Big Bang) origin of the universe etc. It is the general characteristics of polar opposites that when extended beyond certain limit they either switch position by turning into the opposite or become nonsense. It is almost hilarious to see how "scientific" materialism and its opposite, i.e., idealism and modern positivism (Machian, Neo-Berkeleyan, Neo-Kantian etc.) flip-flops on the major questions of natural science. While "materialist" biologists like Richard Dawkins are waging a ruthless crusade against religion and God (and also against dialectical biologists); his fellow scientists in theoretical physics, astrophysics and cosmology are turning to theology and to the same very God Himself!! This is the intractable quandary that anti-dialectical official natural science has led itself into.

Their success in the narrow field of mechanics of macroscopic bodies and an impressive technology based on it gives them the arrogance to claim the invincibility of their empiricism and the absolute validity of the simplistic reductionism of their mathematics based theories, e.g., Big Bang creation of the universe, or a Theory of Everything. They reluctantly accept Darwinism and the principle of evolution, but then take revenge for this little concession to dialectics, by reducing all life processes to the deterministic, simplistic, and mechanistic processes of genes and other bio-molecules only; and opportunistically use the theory of evolution as a shiny armour for their phony fight against the God of theology. On the questions of society and history these "materialists" turn 180 degrees to the opposite, asserting that no theories or generalized concepts are applicable in these fields; these are absolutely chaotic, random; everything depends on personal whim, caprice, or spurious local causes. History for them is a mere playground of kings, heroes and beauties etc.

The "materialist" philistines who want and wait for "objective reality" to "mature", manifest and develop by itself into a fixed, unchanging category for them to try to "understand" it and who will not act to change their circumstances can never appreciate the dialectical materialism of Marx. The proposition in Theses on Feuerbach that "The philosophers have *interpreted* the world in various ways, the point however is to *change* it" is not only a vast improvement over passive and metaphysical materialism but this also represent a profound and definitive advancement of dialectics over that of Hegel. One *knows* things by *changing* them or by observing them in their past and present change, in the process of their interconnectedness in Nature and in the qualitative leaps they undergo during these changes. Mere contemplation by gazing at isolated, apparently static, dead, or finished things can lead only to empty and impotent "opinion or theories", but no positive knowledge.

The dialectics of Hegel and its subsequent refinement to the materialist version by Karl Marx, Frederick Engels (1820-1895 A.D.) and V.I. Lenin (1870-1924) provided, for the first time a way for a systematic and a comprehensive understanding of the world, history and man at least in their main basic features in all domains of knowledge and also provided the framework for conscious efforts to further improve the condition of humanity by making the natural and social laws work towards desirable goals.

Now when the monopoly capitalist class rule of society has developed to the highest and moribund stage; when the productive forces yet again find progressively less and less avenue for further development, when medieval bigotry of race, color, religion and tribalism are again raising their ugly head on international scale; at a time, when regressive monopoly capitalism is dragging the world working class back to serfdom of the medieval past and have inflicted an unprecedented human and environmental degradation around the world; and more importantly, when the concepts of atmospheric ozone depletion, global warming and the possibility of the bankruptcy of the capitalist system etc. is entering even into the dumb brain of the greediest of the capitalists; the imperative for the transition to socialist mode of production has become an urgent necessity in the most developed economies of the world. The need for a dialectical approach and the role for "subjective will" in understanding and solving problems on a global scale is pushing itself to the forefront, as is also the necessity for a planned development of man himself and his surrounding.

But as history shows, a class society always engenders the need of a metaphysical world outlook and the notion of a supernatural power to preserve the authority of its ruling class. As the last class-based socio-economic formation, world monopoly capitalism still continues it's tenuous and miserable existence. In the sphere of natural science it has no other recourse now but to harness ideological support by creating phantoms only at the unknown reaches of the cosmos and in the nebulous realm of mathematical idealism.

In a realm where positive knowledge is sparse, creators of phantoms of course have a natural advantage. This is especially true when they retain absolute monopoly on the means and the tools for collecting facts, suppress the facts that are not to their liking and choose to collect selectively only those facts that re-enforce their pet phantoms. The two dialectically opposed world outlooks, vis-à-vis the "the view of understanding" and "the view of dialectics" would necessarily lead us to exactly opposite conclusions and programme for actions. With the decline and discredit of theology, monopoly capitalism has now hired natural science to preach theology, in order to intimidating the working people by invoking the fantastic notions of the creation of the universe in a single cataclysmic act of the Big Bang and other mysteries. For dialectics there is no such giant leap or mysteries in the world, because the world is made up entirely of leaps!!

To help facilitate the complete liberation of mankind from class rule, it remains one of the most important tasks of dialectical materialism to undo the last castle of God erected by modern official theoretical natural science and cosmology; by projecting its own scientific insights, guided by its successes in the realm of terrestrial Nature. Above all, the question that now confronts us in cosmology is, what in the final analysis is the future prospect of life and consciousness in this universe – a question that resides at the core of the heart of us all? Monopoly capitalism has usurped absolute authority in the study of the cosmos and in projecting its narrative to reinforce the obscurantist ideas of theology.

Predictably enough, our evangelical official cosmology provides a grim prognosis –a sort of doomsday indeed:- depending on some measurable parameters, the universe will either collapse back to the size of a proton crushing us and everything else with it, thus reversing the process of creation (this cycle may be repeated if the creator so wishes), or it will keep on expanding

21

(at predictable rates) with the helpless humans watching in horror as they are torn away from their familiar neighbouring galaxies and life will be lost for ever in a continually-expanding space-time etc. etc. This impels us to bow our head in all humility and reverence not only to the creator, but also to the experts who are in the know of His cosmic design!

In the words of non-conformed and now exiled prominent astronomer and astrophysicist Halton (Chip) Arp: (Personal communication) "Monopoly capitalism vs. desire for knowledge and curiosity about how we and the world works. Fear vs. courage. Is our historical base line long enough to know which will prevail? The galaxies and quasars make me somewhat hopeful"

And most of all, one should always remember the passionate and dialectical assessment of Frederick Engels (Dialectics of Nature):

"… however many millions of suns and earths may arise and pass away, however long it may last before the conditions for organic life develop, however innumerable the organic beings that have to arise and pass away before animals with a brain capable of thought are developed from their midst, and for a short span of time find conditions suitable for life, only to be exterminated later without mercy, we have the certainty that matter remains eternally the same in all its transformations, that none of its attributes can ever be lost, and therefore also, that with the same iron necessity that it will exterminate on the earth its highest creation, the thinking mind, it must somewhere else and at another time again produce it."

The Infinite – As a Hegelian Philosophical Category and Its Implication for Modern Theoretical Natural Science*

Abstract: The concept of the infinite as a mathematical, a scientific and as a philosophical category is differentiated. A distinction between Hegel's dialectical concept of the infinite as opposed to his idealist-philosophical "system" of the "Absolute Idea" as the "True Infinite" is emphasized.

A) As a Mathematical Category

The concept of the infinite as a mathematical category arose naturally enough with the invention of the numerical system by the Sumerians around 3000 B.C. and the subsequent developments of the concepts of geometry, the measure of time, mathematical operations (arithmetic, algebraic, exponentials etc.), One could always add or subtract a unit of number, length or time to get a new one ad infinitum without an end. This infinite is undetermined, has no characterization and was termed the "spurious" or the "false" infinite (*bad infinity*) by G.W.F. Hegel (1770 – 1831 A.D.), as opposed to the "True Infinite" (to be discussed later).

"The spurious infinite" according to Hegel [1], "…seems to superficial reflection something very grand, the greatest possible. … When time and space for example are spoken of as infinite, it is in the first place the infinite progression on which our thoughts fasten … the infinity of which has formed the theme of barren declamation to astronomers with a talent for edification. In an attempt to contemplate such an infinite our thought, we are commonly informed, must sink exhausted. It is true indeed that we must abandon the unending contemplation, not however because the occupation is too sublime, but because it is too tedious … the same thing is constantly recurring. We lay down a limit: then pass it: next we have a limit once more, and so for ever."

The infinite as a mathematical category took a mystical form once Pythagoras of Samoa (580(?) – 520 B.C.), and later Plato (429 – 347 B.C.) idealized the numbers, their relations and geometry into their philosophical system, where the infinite along with the numbers and the forms were universals that exists in a realm beyond space and time for all eternity, a realm that sense perception cannot reach; it is only given to thought and intuition.

As Frederick Engels [2] wrote,

"Like all other sciences, mathematics arose out of the *need* of man; from measurement of land and of the content of vessels, from computation of time & mechanics. But, as in every department of thought, at a certain stage of development, the laws abstracted from the real world become divorced from the real world and are set over against it as something independent, as laws coming from outside to which the world has to conform. This took place in society and in the state, and in this way, and not otherwise, *pure* mathematics is subsequently *applied* to the world, although it is borrowed from this same world and only represents one section of its forms of interconnection – and it is only just precisely because of this that it can be applied at all".

The mathematical pursuit of the infinite therefore, of necessity became a spiritual endeavor. In his attempt to know the infinite and to prove his continuum hypothesis, Georg Cantor (1845 – 1914 A.D.) for example, was eventually compelled to make a distinction between *consistent* and *inconsistent* collections; for him only the former were *sets*. Cantor called the *inconsistent* collections *the absolute infinite* that God alone could know. His idea of an "actual infinite" attracted theological interest because of its implication for an all-encompassing God; but at the same time it inspired scorn of the contemporary mathematicians. What Cantor, other mathematicians and natural science pursued in reality is the "spurious infinite" of Hegel. An infinite series starting with a first term is also undefined, because there is no end to the other side, and one cannot come back to the first term starting from the other end. Cantor's pursuit of the infinite led him to the ridiculous idea of the "*infinity* of infinities, and no other mathematicians followed his steps. If there is more than one infinite then by definition they become mere finites. Mathematicians of all ages had no clue as to the nature of the infinite; some denied its existence all together; while others maintained (following Plato) that mathematical entities cannot be reduced to logical propositions, originating instead in the intuitions of the mind.

B) As a Scientific Category:

Historically, natural science took a rather pragmatic and an opportunistic approach towards infinity, i.e., *reductio ad absurdum* argument which avoids the use of the infinite. It truncates infinity by putting an arbitrary limit as Georg Cantor did, and calls the rest the "absolute infinite" that is known only to infinite God. It deals with infinity with some arbitrary mathematical tricks, for example, a circle is the limit of regular polygons as the number of sides goes to infinity; an infinite series starts with a first term; in renormalization, one set of infinite is cancelled by invoking another set of infinite to get a finite result that was desired in the first place and so on.

Isaac Newton (1642 – 1727 A.D.) and Albert Einstein (1879 – 1955 A.D.) faced the same conceptual problems of the infinite universe in formulating their theories of gravity. Einstein declared, "Only the closed ness of the universe can get rid of this dilemma" [3]. He then set himself to develop a theory of gravity based on geometry, because geometry deals with closed space!

But an attempt to truncate infinity this way can only lead us back to medieval geocentric cosmology. The unpleasant fact is that, by definition a truncated infinite is also infinity and any mathematical operation on infinity leaves it unchanged as Galileo asserted in his famous 1638 pronouncement on infinity that, "Equal", "greater", and "less" cannot apply to infinite quantities [4]. The arbitrary renormalization process and *reductio ad absurdum* practiced by natural science cannot resolve the contradiction of the infinite; it only leads to more and more contradictions and a dependence on ever more mysteries and theology, as we observe in modern theoretical natural science. The reason why Albert Einstein chose a finite and closed universe as opposed to the open ones was not only to make his equations meaningful and/or because of his love for *simplicity* and *aesthetics*, as reductionist ideologues and worshipers of symmetry would have us believe, but also because of his sober realization that his Machean-philosophy based cosmology collapses in an infinite universe. If Mach's principle is followed, then an infinite universe means that the inertia and the mass of atoms etc. also become infinite. To keep the world as we see it now (inertia, mass, etc.); all Mach based cosmologies must have the universe started at a finite past and also must have a finite extension. So this way the contradiction of infinity is not solved.

The notion of the infinite in natural science became ever more clouded after Albert Einstein established the primary role of mathematics in natural science. Natural science became seduced to the idea that where experimental evidence and empirical data is difficult and/or impossible to obtain "logical consistency of mathematics" will lead the way. The stunning success of the theories of relativity in early 20th century, led Einstein to revive Pythagoras's notion of mathematics. "How can it be" he wondered, "that mathematics being a *product of human thought* which is *independent of experience*, is so admirably appropriate to the objects of reality?"[5]

The theory of general relativity is a classic example where the power of mathematics, pure thought and aesthetics devoid of any

26

empirical content is purported to have conceived the ultimate reality of the universe. "Our experience hitherto justifies us in believing that nature is the realization of the simplest conceivable mathematical ideas. I am convinced that we can discover by means of purely mathematical constructions the concepts and the laws connecting them with each other, which furnish the key to the understanding of natural phenomena. … In a certain sense, therefore, I hold it true that pure thought can grasp reality, as the ancients dreamed", declares Albert Einstein [6].

With his mathematical idealism Einstein erased the difference between the *pure* mathematics, whose program is the *exact* deduction of consequences from logically independent postulates, and the *applied* mathematics of *approximation* needed for science. Natural science uses approximate empirical data, which are fitted on in various ways to *analytic functions* of *pure* mathematics that helps in the systematization, generalization, and the formulation of tentative theories. But the results and the inferences are only valid in a narrow range of the data values for the argument for which approximate empirical information is available.

A convenient property of the analytic functions (such as the field equations) is that, such functions are known for all values of their argument when their values in any small range of the argument values are known and thereby allowing an unlimited extension of this procedure from the macrocosm to the microcosm. Thus, the *a priori* assumption that the laws of Nature involve *analytic functions* leads to a complete mechanistic determination of the world based on their experimentally determined value in a narrow range only. But the validity of such a procedure of unlimited extension of mathematical functions for the real world, were questioned both by mathematician/philosophers such as Bridgman [7] and scientists like Klein [8] at the advent of quantum mechanics; based as they argued (on different grounds) on the unavoidable inaccuracies of empirical knowledge. And as quantum mechanics clearly shows,

27

there is uncertainty in the ontological nature of reality itself at micro level. So, our epistemological knowledge must always be defective, tentative and approximate, increasing in scope from one generation of humanity to the next; like an infinite mathematical series, without ever coming to a termination or without ever reaching one final and ultimate truth.

The quantum phenomena and the failure so far (9); (in spite of over a century-long intense efforts by some of the most brilliant mathematicians including Einstein) to unify "ALL" the particles and "ALL" the forces of Nature into a simple and reductionistic "theory of everything" demonstrate the folly of this kind of naïve and over- simplified extrapolation of idealized mathematics to the real world at the two opposite directions of infinity, i.e., macrocosm and microcosm.

C) As A Philosophical Category

The concept of the infinite was implicit in the early philosophical developments especially among the early Greek thinkers that centered around the basic questions of the primacy of spirit or nature, unity or multipliticity, stasis or motion. This debate divided the philosophers into two great camps. Those who asserted the primacy of spirit, unity and stasis formed the camp of idealism; the contrary camp formed the various schools of materialism.

The earliest idealist Greek philosophers (the Eleatics) denied the reality of becoming, multiplicity or motion; these characteristics they maintained, are of the sense-world or physical Nature. These they argued are not *real* but only *appearances* and hence these are illusions. For Parmenides (515 – 450 B.C.) for example the sole reality is Being, Being is One, only the One is; the Many not. This Being cannot be perceived by senses, it is given only to thought or mind. This line of thinking permeates the range of idealist philosophers like Plato, Aristotle, Berkeley, Hume, Hegel and all monotheistic religions. The Unity of Being in this view means that the infinite must be contained in this one Being.

28

The Being meaning God in theological terms, the infinite, then became associated with abstract God. The idealist view of infinity was later incorporated into mathematics and theoretical natural science.

But the dialectically opposite and the materialist view of reality – i.e. the validity of the sense perception of change, multiplicity and motion in material Nature also arose simultaneously in early Greek philosophy. The founder of the dialectical view, Heraclitus (544 – 483 B.C.) on the contrary saw the world as a process – as changing eternally. For him Unity is not a homogenous unity, but is a "unity of the opposites or of opposite tendencies". The Unity is a complex entity that contains at least two dominant opposite fragments that are in constant conflict with each other and renders this unity susceptible to diversity, change and movement. The concept of the infinite in this view is therefore, open ended. Epicurus (341~270 B.C.) following the tradition of Heraclitus was the first to assert that the universe is infinite in its extension in all directions and that multiplicity, time and motion are endless.

Benedict Spinoza (1632-1677 A.D.) made an important advance on the concept of infinity along the dialectical tradition which helped Hegel (himself an idealist) to formulate in a comprehensive way the dialectical view of the infinite in particular and his dialectical method in general. Spinoza formulated the profound idea that to define something is to set boundaries for it; i.e., to determine is to limit. The infinite then is something that is undetermined or that has no limit or boundary. In other words the Infinite is limited only by itself and like God is "self-determined".

In popular concept, God is supposed to be infinite. Spinoza's idea of the infinite led to an insurmountable difficulty for conventional philosophy and theology which regarded the infinite and the finite as mutually exclusive opposites; absolutely cut off from each other. How then the infinite can be conceived; how infinite God can have contact with finite man, since it will

29

limit His infiniteness. Finiteness of the world became a primary requirement for medieval theology. The inquisition did not hesitate to spill blood and torture victims to defend its doctrine. Hegel, following Spinoza called the "Absolute Idea" of his philosophy the "True Infinite" which is self-determined. For him the material world or Nature is a crude replica – an alienated form of the "Absolute Idea".

The fundamental difference between these two worldviews and hence their implication for the concept of infinity gets its concrete expression in the question of *matter* and *motion*. While Newton recognized matter as a *real entity*, for Einstein matter is a particular representation of an all pervading (space-time) reality ("Being" of Parmenides?). Einstein expressed this point of view in an unambiguous way, "Since the theory of general relativity (GR) implies the representation of physical reality by a continuous field, the concept of particles and material points cannot play a fundamental part and neither can the concept of motion. The particle can only appear as a limited region in space in which the field strength or energy density is particularly high" [10]. Motion in the view of both Newton and Einstein could only arise from an *impulse* from without - from God – the "unmoved mover". And why energy density at particular points must arbitrarily be high to form material points must also depend on intervention by Providence. For dialectics (and quantum mechanics) on the contrary, matter and motion are the fundamental elements and the primary conditions of all physical reality; *motion is the mode of existence of matter. Matter without motion is as inconceivable as motion without matter.*

The only way the conceptual problem of infinity can be resolved is through the dialectics of Hegel - the law of the unity of the opposites. The notion that the finite and the infinite reside together in a contradiction; that they are united as well as are in opposition to each other. That, the finite **is** the infinite and vice versa. That this contradiction resolves itself continuously in the never-ending development in time and extension in space of the universe, in the same way as for example intellectual advance find its resolution in the progressive evolution of humanity from one particular generation to the next. Just as Nature or the universe (ontologically) is incapable of reaching a final, ever lasting, unchanging or an ideal state so is thought (which is only a reflection of Nature in the mind of man) epistemologically is incapable of comprehending a completed, exhaustive or immutable knowledge - the so-called absolute truth of the world. For dialectics, "eternal change" (with temporary stages of infinite number of leaps) is the only thing that is permanent and the only absolute. Hegel's dialectics therefore, is a condemnation of all claims to absolute truth by all idealism including the mathematical idealism of modern official natural science, which is but a reincarnation or rather restoration of the old idealism. In human history, as well as in the history of natural science, hitherto all claims to the "final truth" are but the partial masquerading as the complete.

The continuous resolution of the contradiction of the finite and the infinite like the other evolutionary processes are not only dialectical but they also develop historically following the three general laws i.e. i) transformation of quantity into quality and vice versa, ii) interpenetration of the opposites and iii) the negation of the negation. Engels [11] summarized these three laws from Hegel's *Logic*, where the first law comprises the *Doctrine of Being*, the second, the *Doctrine of Essence*, while the third constitutes the fundamental law for the construction of the whole system. Hegel deduced his philosophy from the history of Nature, of society and of thought. The infinite universe is not a mere abstract, quality less, boring, endless extension of uniformity (spurious or *bad infinity*), it includes a variety of qualitative contents with different forms of movements passing one into the other and developing historically. The infinite space is adorned with the drama of things "coming into being" and "passing out of existence" in each of the innumerable island universes; each island universe with innumerable galaxies and each galaxy in turn with innumerable stars and planets. Under favorable conditions, galaxies propagate [12, 13]; the stars produce the higher elements; the planets give rise to the evolution of molecules, to organic life and finally to the thinking brain through which infinite Nature (for a brief period of time) *becomes conscious of itself*! Self-consciousness is therefore, the property of the highest developed form of matter, which like everything else comes into being and passes out of existence as temporary bubbles in the eternal and infinite universe.

The knowledge of the infinite is therefore proportional to the knowledge of the finite. This knowledge is necessarily a historical and an iterative process progressing through successive generations of mankind without ever terminating in one final or absolute truth a quest of which was the aim of all idealism – mathematical, scientific or philosophical. A progressively better understanding of the infinite universe can only come about by studying the finite around us guided by the general laws of dialectics.

There are infinite number of water and other molecules and atoms on earth and yet we understand (in a limited sense) and live at ease with these! The properties of matter and its structure under the various conditions in terrestrial nature must be the same that exists under similar conditions billions of light years away. In fact, one sun with its planets and its life supporting earth and one Milky Way galaxy with its surrounding family group form the essential basis for an understanding of the universe. Beyond 15 billion light years there is no wonderland or lurking monsters to be seen. What we will see there is more or less the same we now see within a few million light years around us! The same applies to the micro-world. There is no limit of space, time or length in any direction; up-down, left -right; back – front, at least up to the level beyond which the terms mass, time or length lose their meaning (in the usual sense of the term) because of quantum uncertainty and due to other yet unknown effects. The limits from quasars (at the ultimate boundary of the universe?) to the quarks at the lowest end, set by Official Science must therefore be false; because this represents an arbitrary limitation of infinity, conditioned by the limitation of the empirical knowledge of our time.

E) The "Absolute Idea" of Hegel – as the "True Infinite"

As Engels pointed out [14], the dialectical view of the infinite as discussed above, are necessary logical conclusions from the dialectical method of Hegel; but conclusions he himself never expressed so explicitly. Hegel was an idealist and above all he was the official philosopher of the Royal Prussian court of Frederick William III. His task was to make a system of philosophy that must specify one absolute truth or a "first cause" of the world, as tradition demanded it. Therefore, even though Hegel, especially in his *Logic* emphasized that this absolute truth is nothing but the logical. i.e., historical *process* itself, he nevertheless found it necessary to bring his dialectical process to a termination in the "Absolute Idea". For his philosophical "system" his dialectical "method" had to be untrue. Hegel also turned his philosophy upside down, where the "Absolute Idea" (like all idealism) became primary and nature only a crude reflection of the "Idea", even though (through unprecedented detail and encyclopedic work) he extracted the laws of dialectics from the history of the material and the human world.

But nevertheless, the dialectical method of Hegel helped him to overcome the impossible contradiction of the infinite and the finite faced by Spinoza, theology and all previous idealist philosophies. For Hegel, the finite and the infinite are no independent entities separated from each other by an unbridgeable gap in between, as old philosophy asserted; but these are the integral components of a single unity within which the two opposites reside together in active unity and opposition, and hence in a logical contradiction. A resolution of this contradiction to an ever new "unity of the opposites" and so on – *the negation of the negation* is what gives rise to motion, change, development, and historical evolution of the universe as a never ending process.

Idealist Hegel can terminate the infinite process of change by making his "Absolute Idea" (the self determined, the True

Infinite") as the ultimate end result of all change, motion, development or history, and making it the beginning again, i.e. the end as the true beginning. For Hegel, the finite Nature or man IS the infinite "Absolute Idea" itself! The "Absolute Idea" alienates and disguises itself into Nature, evolves historically through all the usual twists and turns following the laws of dialectics and comes back to itself again through the consciousness of man and particularly through the philosophy of Hegel himself, who for the first time in the history of mankind perceived in thought the ultimate truth of this dialectical movement, in absolute profoundness. For Hegel the "Absolute Idea" which is the end result of all change, development, motion, history etc. - the static reality of Parmenides, the abstract God of theology, the self-determined entity of Spinoza, is the "True Infinite" and the absolute truth of the world.

But this "Absolute Idea" or the "True Infinite" of Hegel like the mathematical "Absolute Infinite" of Cantor; are only absolutes in the sense that they have absolutely nothing to say about it! Thus in spite of his prodigious intellect and in spite of the logical implication of his profound dialectical "method" to the contrary, Hegel unfortunately pursued the illusion of an absolute truth, like all the other idealist philosophers and all theological prophets of all times. The mathematical idealism and reductionism of modern official theoretical natural science inherited this illusion – i.e., the empty shell of all idealism but not the kernel - the dialectical "method" of this great idealist thinker.

E) Conclusion:

During the last few centuries especially since Copernicus (1473 – 1543), natural science accumulated impressive empirical evidence and gained variable degrees of understanding of the terrestrial nature; that collectively vindicate Hegel's assertion that *change* is the only absolute truth and that the dialectical laws are the only eternal laws that govern the development and the transformation of matter and life. But ironically, natural science claims its own invariable truth exactly in the areas where it possesses the least empirical evidence!

As intoxicated modern official natural science celebrates its achievement of a definitive knowledge of one single event i.e., the "Big Bang" origin of the universe and the triumph of its mathematical idealism; with the award of Nobel Prizes, and as the world awaits in breathless anticipation the imminent discovery of a "theory of everything" that will bring an "End of Physics" and possibly the end of all knowledge (by "knowing the mind of God", according to one of the leading physicists Stephen Hawking [15]); it would be instructive for us to remember the sober dialectical assessment of Frederick Engels [16] - one of the greatest inheritors of Hegel's philosophy:

"The perception that all the phenomena of Nature are systematically interconnected drives science to prove this interconnection throughout, both in general and in detail. But an adequate, exhaustive scientific statement of this interconnection, the formulation in thought of an exact picture of the world system in which we live, is impossible for us, and will always remain impossible. If at any time in the evolution of mankind such a final, conclusive system of the interconnections within the world - physical as well as mental and historical – were brought to completion, this would mean that human knowledge had reached its limit, and, from the moment when society had been brought into accord with that system, further historical evolution would be cut short – which would be an absurd idea, pure nonsense. Mankind therefore finds itself faced with a contradiction; on the one hand, it has to gain an exhaustive knowledge of the world system in all in its interrelations; and on the other hand, because of the nature both of man and of the world system, this task can never be completely fulfilled. But this contradiction lies not only in the nature of the two factors – the world, and man – it is also the main lever of all intellectual advance, and finds its solution

35

continuously, day by day, in the endless progressive evolution of humanity…".

F) References:

1) Wallace, W. Trans. *The Logic of Hegel*, Oxford, Clarendon Press, 1892, §94 Z., Cited in *The Philosophy of Hegel* By Stace, W.T., N.Y. Dover, 1955, § 198.

2) Engels, F., *Anti-Dühring*, N.Y., International Publishers, 1939, pp. 46.

3) Kragh H., Cosmology & *Controversy*, Princeton Univ. Press, 1996, pp.07

4) Kaplan, R., and Kaplan, E., *The Art of the Infinite*, Oxford Univ. Press, 2003, pp. 228.

5) Einstein, A., *Sidelights on Relativity*, N.Y., Dover, 1983, pp. 28.

6). Einstein, A., *Essays in Science*, Trans. Alan Harris from "Mein Weltbild", Quedro Verlag, Amsterdam, 1933, N.Y., The Wisdom Library, 1934, pp.16 – 17.

7) Bridgman, P.W. *The logic of Modern Physics*, N.Y. The Macmillan Co. 1927.

8) Klein, F. *Elementarmathematik von Höheren Standpunkt aus*, Berlin, Vol. 3, 1924. Cited in *Quantum Mechanics*, by E.U. Condon & P.M Morse,., N.Y. McGraw-Hill Co. Inc., 1929, pp11.

9) Smolin, L. *The Trouble with Physics*, Boston, N.Y. Houghton Miffin Co., 2006.

10) Einstein, A. *On the General Theory of Relativity*, in David Levy (Ed.). *The Scientific American Book of the Cosmos*, N.Y., 2000, pp. 13.

11) Engels, F. *Dialectics of Nature*, N.Y. International Publishers, 1940, pp. 26.

12) Malek, A. 'Ambartsumian, Arp and the Breeding Galaxies', Montreal, *Apeiron*, vol. 12, no.2, 2005, pp. 256-271.

13) Arp, H.C., *Seeing Red: Redshifts, Cosmology and Academic Science*, Montreal, Apeiron, 1998.

14) Engels, F. *Ludwig Feuerbach and Outcome of Classical German Philosophy*, N.Y. International Publishers, 1941, pp.12-13.

15) Hawking, S. *A Brief History of Time*, N.Y., Bantam Books, 1990, pp. 175.

16) Engels, F. *Anti-Dühring*, N.Y. International Publishers, 1939, pp. 43-44.

*This article first appeared in the Internet based journal Mukto-Mona on 9 February 2009

Ambartsumian, Arp and the Breeding Galaxies*

Abstract: Two exactly opposite views about the origin, the evolution and the formation of galaxies in the universe are discussed. The first one, which is mainly based on mathematical idealism and is generally accepted; views galaxy formation as deterministic and an essentially unidirectional condensation of diffuse matter created through a single primordial explosion (The Big Bang) about fifteen billion years ago. The second view, based on (limited) observational and empirical evidence asserts a rather intrinsic origin of galaxies, where new galaxies are formed from material ejected and/or dissipated from the core of existing galaxies. A dialectical perspective in support of the second view is presented.

Keywords: Galaxies, "Big Bang" theory, dialectics, condensation, primordial gas, ejection/dissipation, gravitational relaxation, appearance/disappearance, annihilation, matter/antimatter.

One major characteristic of an invariable truth (scientific truths included) is that, it makes its appearance at a certain time in history, but nevertheless it lays claim to unconditional validity for the past, the present and the future. Once it is extracted from the real world, it becomes an alienated subjective force and assumes an independent entity of its own as if coming from outside. It then sets itself to control its very creators—religion, state, capital are a few examples. Another important aspect of such a truth is that it has its origin in a "first cause"—a mystery, which has to be accepted as an article of faith; the rest follows from it in a deterministic way, obeying the laws of cause and effect. Any new fact or phenomena must fit in the scheme of this truth; if it fails to do so then secondary, tertiary, *etc.*

38

mysteries have to be incorporated to make it fit in the larger scheme.

The General Theory of Relativity is such an invariable truth. Matter and space-time is engaged in a sterile and eternal love embrace, and this describes the architecture and the geometry of the universe. Albert Einstein himself claimed that one can only prove or disprove this theory, but any further improvement of it is impossible. The world then follows from this truth obeying the laws of causality. The Big Bang theory, inflation, dark matter/energy, black holes, *etc.*, are secondary, tertiary mysteries that need to be invoked to bring the new cosmic phenomenon in line with the primary truth.

The Big Bang theory claims to be an invariable truth on its own merit specially, after the rival Steady State theory lost much of its appeal. Both of these theories were proclaimed as corollaries to the General Theory of Relativity. If the Big Bang theory required of God to create the universe at one stroke and then either forget about it or follow on an eternal cycle of Big Bang and Big Crunch; the rival theory obliged Him to keep an inventory and keep on creating matter where and when necessary for all eternity.

According to Big Bang theory, the universe came into being with a primordial explosion no-where and no-when and is destined to follow a predetermined course set out as a mathematical plan. Starting from the size of a proton and undergoing an initial unimaginable rate of inflation, the universe is continuously expanding ever since. All matter/energy, space-time or anything else in this universe including us, are shards of a ten or twenty-six dimensional reality of Plato's ethereal realm of perfect symmetry, exquisite beauty and absolute order. This ethereal realm is not given to our senses. We can reach it only through the power of thought and by following the logical consistency of mathematics. We can possibly get a glimpse of the original reality by piecing together the shards that are strewn around us in this universe and putting them in their place in the puzzle. The task of physics and cosmology is to reveal the image of the original reality in the details of the cosmos. Dark matter/energy, black holes, *etc.* are mini-mysteries needed to explain the dynamics of the galaxies. After completing the "theory of everything" (which is not far off) we will "know the mind of God," and can live happily ever after.

The textbooks and professional articles on astrophysics, astronomy, cosmology, *etc.* start their deliberation assuming Big Bang theory as granted. Any meaningful or at least rewarding research, studies, *etc.* in this area must be concerned in finding the glory of this truth in heaven and nature. At stake are generous research funds, lucrative positions and most of all instant fame & glory. Only positive results are worthy of publication or discourse, the negative ones are of no importance or consequence at all. In the jungle of electronic noise in the spectra or pictures, one must hunt for positive signs of this truth.

There is no doubt that controversies, debates, *etc.* exist regarding the cosmos and volumes are written and spoken. But all these are involved with what happened *after* all matter/energy burst forth from the single act of creation, or on the precision of the measurement of certain cosmological parameters, or the formulation of various mathematical models, *etc.*, *etc.* But there is absolutely no doubt that the galaxies condensed from the fixed amount of gas formed during the act of the creation. "Fierce" debate is raging on whether the gas cloud broke into huge chunks, which later fragmented to form stars, galaxies, cluster of galaxies (top down) or whether smaller chunks of the initial gas cloud condensed first and then grouped together to form galaxies, *etc.* (bottom up). Except for a few minor details, Big Bang cosmology is a satisfactory description of the universe, thereby bringing astrophysics and astronomy to a close. The non-conformists to this paradigm and critics, who undertake investigation of the universe just as it is, or on a different premise, are but "gadflies" who only cause vexing and unnecessary irritations.

Let us briefly recount the history of our invariable truth of the universe and official observational cosmology of the recent past. The invariable truth started in 1916 with an all-inclusive equation. But a solution of this equation showed that the universe should be unstable (expanding or contracting) which was contrary to the conventional perception, so a fudge factor was put into the equation to keep order and peace in the heavens. A proof of this "now invariable" truth was needed and it came soon enough. An experiment in 1920 led by Arthur Eddington (a mystic of numerology) measured the bending of star- light by the gravitational power of the sun; the bending was exactly what was predicted by the theory! But soon that particular experimental proof was also found to be a fudge factor, which Stephen Hawking described as "a case of knowing the result they wanted to get, not an uncommon occurrence in science". Edwin Hubble's discovery in 1929 that the galaxies seem to be flying away from each other at a rate proportional to their distance, also discovered the "the greatest blunder" of a life time, the fudge factor in the original equation was un-necessary after all. Now that we have got back our invariable truth in its original form, an unlimited extrapolation of Hubble's finding must mean that everything in this universe was at one point in the past, from which it started off with a bang, giving the result we see today. The discovery of the microwave background radiation in 1965 *sealed the deal* forever. It showed a *perfect* isotropic picture of the universe that it was supposed to be. But there were minor glitches and irritations from the "gadflies" like Gerard Henri Vaucouleurs, George Abell, Vera Rubin and others— observations revealed that the universe was not homogeneous at all as far one could survey (more than 15% of the supposed universe!) with the most sensitive tools of astrophysics. Instead matter is seen to be clumped progressively into stars, galaxies, groups of galaxies, cluster of galaxies, clouds, super-clusters, super-cluster complexes that can span hundreds of millions of light years and so on, often linked together in filament-like strings that border vast region of empty space where there is almost nothing at all.

Well, not to worry, we can now let quantum mechanics into the picture to create some minor fluctuation in the primordial atom. These fluctuations must survive the super-bang explosion, separation of the various forces, separation of matter and radiation, the incredible super-luminal inflation of the early universe and so on to form the clumpy structure of the cosmos. And one must now look for this in an anisotropy of the cosmic microwave background (CMB) radiation that was already found to be isotropic. Sure enough, experimental evidence came handy in 1977 when George Smoot and co-workers detected the CMB anisotropy in an experiment conducted from a U-2 aircraft. The "final proof" came on April 23, 1992 in the COBE (Cosmic Background Explorer) satellite experiment again under the direction of Smoot (who else?). A lot of drama, tension, expectation prevailed during the two years of data collection. There was a lot at stake! Instead of devising or doing their own experiments, scientists all around the world were all hushed up in breathless anxiety and expectation about what this new Messiah was going to pull out. Some of them were saying, "You could say we're close to a crisis, but the truth is, we're getting down to the point where we should see fluctuations. We are now positioned to see them—and boy, we'd better see them…" [Quoted by T. Ferris, *The Whole Shebang*, Simon & Schuster, 166(1977)]. As expected the result came out positive —sigh of relief, the invariable truth survived yet another big test. It was greeted with comments like, "The scientific discovery of the century—if not all time," "the Holy Grail of cosmology" and so on. COBE must finally put an end to the torture of the gadflies or any doubt as to the absolute truth of the Big Bang! But of course the COBE results had to be positive like Eddington's experiment, amid all this hype, expectations, browbeating, *etc*. A negative result would have necessitated "further" experimentation until we got a positive one. In science as in life, we eventually get what we are looking for, because we already *know* that the thing exists! Nevertheless this episode speaks a lot about a science and its invariable truth when you have to hold your breath on the outcome of a single experiment. It is all the more remarkable that neither the invariable truth of the Theory of Relativity, nor any aspect of astrophysics, cosmology related to Big Bang genesis so far merited for a Nobel Prize.

But what happens if we try to look at the cosmos, stars, galaxies, *etc*. in this universe as they actually are, free of any preconceived idea or without the lens of an invariable truth and then form theories, intuition, *etc.* based on these observed facts? Viktor Ambartsumian tried to do exactly this in the 1950s. He insisted that observation must take precedence over speculation in the study of the cosmos. He had found "stellar associations"—groups of ten to a thousand stars within the Milky Way apparently of a common origin but they were moving too fast for their gravity to hold them together permanently and to stop their slow dispersion. Ambartsumian saw similarities between the dispersion of stellar associations and other cosmic phenomena such as the ejection of matter/energy from dying stars, the gradual break-up of binary stars and later extended it to include the more catastrophic ejections in radio galaxies, quasars, *etc.* in which he found the same dispersive process in action. Ambartsumian suggested that material is dispersed/ejected from the galactic nuclei variously producing the intergalactic gas, the feature such as the spiral arms and/or giving birth to new galaxies or clusters of galaxies. Dispersion from compact (super-dense) sources at the galactic core for him, therefore, represented the fundamental dynamics of the universe. This is exactly the opposite of the view that the galaxies condensed from the diffuse clouds of gas produced from some events in the past as Big Bang and Steady State theories proposed. Ambartsumian of course was humble enough not to theorize about the nature of the compact source or how catastrophic ejection of matter/energy could take place, simply noting that there was no known physical process that could cause such enormous events. He probably did not want to minimize the great significance of these observations and his intuition about them, by invoking some fearful and cheap cosmological beasts or monsters like black holes, dark matter/energy, *etc.*

But Ambartsumian's work received little attention or at best met with dismissing skepticism. The two camps of theorists were beating their war drums and were busy collecting only those evidences in the cosmos that would support their respective invariable truth about the creation of the universe. For both of these camps the study of galaxies was of secondary interest, since it was generally assumed that galaxies were formed by the condensation and contraction of diffuse matter. With the discovery of the microwave background radiation the drum-beating of the Big Bang camp became so loud that it drowned out not only the Steady State camp, but everything else—Ambartsumian included.

In the midst of this deafening sound of the Big Bang, very few non-conformists like Halton Arp were trying to bring back "observation" at the center of astrophysics. Arp was carrying on with his tenuous attempts (opportunity permitting) to study the so-called peculiar galaxies, Active Galactic Nuclei, quasars, *etc.* During his long career in astrophysics, Arp along with others have collected spectacular images of the cosmos that show a dynamic process of violent ejection and dispersion of matter/energy as an important aspect of the universe. They found evidence of systems of galaxies linked by jets of nebulosity showing the track of ejection of objects from a parent source where an ejectum may have totally different red shift from its parent. Even without going into the controversy of the intrinsic nature of red shift and variable mass theory, these images and studies themselves confirm the validity of the great insight of Ambartsumian i.e., the dialectical evolution of the universe. Not only is there clustering of galaxies on the largest scale (which is now reluctantly accepted as a fact), in the small scale also, family like groupings of a high concentration of dwarf irregular galaxies, a host of small dwarf elliptical galaxies, distant globular objects, *etc.* that cluster around large spirals like Milky Way and Andromeda were studied in detail by the Estonian astronomer Jaan Einasto. These groupings, known as the local groups now seem to be a general pattern in the cosmos. The giant elliptical galaxies such as M87, possess a population of globular clusters extending at large distances in space. These observations suggest the intrinsic origin of the satellite galaxies and the globular clusters i.e., ejection from parent galaxies as envisioned by Ambartsumian.

All these tedious and painful efforts of the non-conformists could do is to earn them the honour title of "gadflies". But it seems that these non-conformists should rather be called "fireflies" instead, because they provide the only dim light in this dark cloud from the Big Bang that threatens to become more and more "dark"! Observational astrophysics, which like natural science in general started off as a tool for free and open inquiry of the cosmos, and which made great strides at its onset, now finds itself progressively being appropriated by a monopoly economic power and an interest group that use it in support of their pet theories. The fierce conflict between the paradigms of dialectics and causality that raged in the field of terrestrial nature and which was won decisively by the former with the discovery of the theory of evolution and quantum mechanics; has now shifted to the field of the cosmos.

But how things stand if in our search of the cosmos we follow the dim light of these "fireflies" (rather than the intense "dark" beam of an invariable truth) guided by the insight of Ambartsumian and the views of dialectics. According to dialectics, nothing can exist without its opposite residing in its own element at the same time (the *unity of the opposites*). The impetus for any development, change, evolution—in one word *motion* in a thing or a process is due to the conflict of the ever-present opposites residing in the very elements of the thing or the process itself and not because of an "impulse" from outside, as causality believes. For dialectics *motion is the mode of existence of matter. There can be no matter without motion and no motion without matter.* This motion is mediated by blind chance and iron necessity that is inherent in chance Quantum mechanics also has reached the same conclusion. So, the notion of the existence of a primordial entity of perfect (ten or twenty-six dimensional) symmetry, of absolute order, *etc.* that burst forth into this universe through a single event such as the Big Bang is a myth created by our mind and our mathematics. And so must also be the notion of a creator who pushed the button to trigger the Big Bang explosion, since there was no mechanism within the primordial entity of *absolute order* to trigger itself or to come to itself in the first place. According to dialectics *there is no leap in nature,* precisely because *nature is composed entirely of leaps!*

In the case of terrestrial nature we observe that matter, life, history and thought evolve through a series of revolutionary changes (qualitative leaps) according to the dialectical law of *the negation of the negation* or a triad of thesis—antithesis—synthesis mediated by chance and necessity; and brought forth through the conflict of the opposites or the contradiction of heredity and adaptation in its very own units. Chance is blind only when it is not realized in a necessity. If a seed from a plant falls on a stone or by chance carried to the moon, it will not grow there, because there is no necessity for it i.e., no scope of its further development; so this chance is sterile and things end there. But when a chance brings the same seed into a fertile soil, it develops due to the exacerbation of the conflict of the opposites within the seed, it *negates* itself into a plant, which in turn negates itself (*the negation of the negation*) to give an increased quantity of the seed itself. All change, motion, development in this view proceeds through *nodal points* or leaps (governed by specific laws) where dialectical opposites either mutually annihilate each other or are sublated (*aufheben*) into a new synthesis and so on (*the negation of the negation*) and where changes in quantity leads to a qualitative change and vice versa. It is the task of natural science to discover these specific laws and not to impose laws on nature created in the brain of man.

In the realm of the cosmos too, the galaxies (the units in the cosmos) must develop dialectically due to the conflict of the opposites residing in themselves and mediated by chance and necessity. But while in the case of terrestrial nature a definite proportion and amount of atoms were given from which everything else evolved (within a relatively short time in cosmic scale) new matter has to come into being for the proliferation of the galaxies and the universe itself! But how could new matter come forth and how could such enormous amount of energy as manifested in the quasars for example, be generated? The "compact source" of Ambartsumian and the "white hole" at the core of the galaxies suggested by Arp fall into the same trap of causality like the Big Bang! These must have come into being and scattered around in the cosmos through some casual relationship originating in a "first cause" like the Big Bang. Very large-scale matter-antimatter annihilation processes are the only possible known sources of energy that can trigger the catastrophic events like quasars. But none of the theories and hypotheses mentioned above (including Big Bang) allows such a possibility.

A dialectical view of the universe as proposed recently (Apeiron, Vol 10, No. 2. 165-173(2003)) can provide a plausible basis for an understanding of the evolution of the galaxies in particular and the phenomenology of the cosmos in general. According to this view, matter in the form of elementary particles comes into being and passes out of existence (with a finite amount being present at any particular time) as a dialectical and quantum mechanical necessity in the universe, which is void and infinite in space and time. Persuasive evidence from quantum electrodynamics suggests that virtual particles inhabit empty space with an increasing concentration close to an atomic nucleus. Some of these virtual particles can become real (and the real pass back to virtual) as chance events and necessities, by tunnelling effects, and/or as pair production by quantum fluctuation in the vacuum and so on, to give rise to both matter and antimatter. Out of the innumerable possibilities, the law of chance and necessity determines which particles eventually prevail. Chance accumulation of matter and/or antimatter at certain points can then provide the seeds for further growth and development of galaxies, following physical laws. Since the appearance/disappearance of matter is enhanced where mass concentration is high, the galactic centers form the most active sites where new matter accumulates and these centers become the theatre where other random and periodic cosmic events can manifest themselves, such as those that we see as the Active Galactic Nuclei (AGNs), quasars, *etc*. This basic process then can form the fundamental dynamics through which the universe evolves.

Everything in this universe from the galaxies to man are therefore, temporary and dynamic structures governed by physical laws and by the interplay of chance and necessity. Chance accumulation of matter at certain points in the infinite space and their evolution can give rise to the island universe around us, because new antimatter that forms is annihilated by reaction with existing matter giving gamma rays. These gamma rays along with other sources of energy degrade through their interaction with matter to give a quantum mechanically necessary zero-point energy—the observed cosmic microwave background radiation. The chance accumulation of antimatter in variable quantities and its inevitable interaction with matter can explain the origin of gamma and X-rays in the galaxies, the gamma ray bursts, active galactic nuclei, quasars, and other catastrophic events in the cosmos, for which no other source for the outpouring of such enormous amount of matter/energy is known.

Also such large scale catastrophic reaction of matter and antimatter can provide the energy necessary to eject the globular clusters, or large chunks of matter from the mother galaxy that later either grow on their own and/or accumulate by gravitational attraction to form satellite galaxies, *etc*. The Seyfert Galaxy NGC 7603 and its smaller companion provide a dramatic example of such a possibility. The minor axis of the flat spiral galaxies provides the most convenient route along which matter/energy can be ejected out in opposite directions. More limited scale of these random and catastrophic events may lead to the less dramatic effects such as the deformation or limited fission/elongation of symmetrical galaxies to form barred structures.

The morphology of the various galaxies, such as spirals, ellipticals, irregulars *etc.,* must be looked at as the various stages, following one another (and not existing alongside of each other since their simultaneous creation, as official cosmology proposes) in the process of their evolution, dissolution or their transformation into each other; determined at any particular time by the net result of the random processes of appearance/disappearance or matter/antimatter production/annihilation, the collision/interaction of the galaxies with each other, ejections from the core *etc.*, which are governed by chance and necessity and the physical laws.

As suggested by Arp, it is possible that linear ejections of pairs of objects along the minor axis of the parent galaxy give rise to the companion galaxies, quasars, *etc.*; while ejections along the major axis are stopped close to the ejecting parent which then may form the spiral arms as Ambartsumian envisioned. Evidence for such a transformation had already been found in 1961 when the French astronomer G. Courtès discovered that proto-spiral arms seem to have been ejected from the center of the Seyfert galaxy NGC4258. A more recent NASA image of this galaxy is shown below:

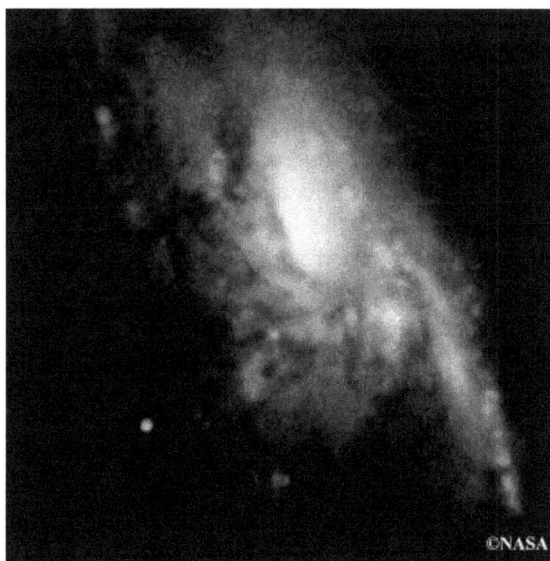

It is entirely possible to speculate that some of the ellipticals transformed into spirals by forming proto-arms through ejection from the core, because unlike the case with the spirals, ejection along any direction will be slowed down by the existing bulk of matter. The spirals then slowly convert back to the SO or E type galaxies through gravitational relaxation. The different types of galaxies are dynamic structures in the various stages of formation/ dissolution, succession/inter-conversion to each other (and not structures "perfect in themselves" since their formation after the Big Bang). It is however, obvious that the shapes of some of the galaxies were determined by the random collision and/or close encounter with another galaxy

The morphology of the galaxies is therefore, mediated by the dialectical process of dispersion/ejection/deformation initiated by the catastrophic events of matter/antimatter annihilation and the regularizing effects of gravitational attraction.

Time is an intrinsic and a relative characteristic parameter for a particular particle or a unit of assembly. It begins when the unit comes into being and ends with its passing away out of existence or with the dissolution of the unit. This view is the exact opposite of the notion of the creation of space-time and all matter/energy in the finite past. If stars, galaxies like humans are temporary entities, "coming into being and passing out of existence" as dialectics asserts, then we cannot measure the age of the universe at any particular moment by measuring the lifetime of the longest living star or galaxy, anymore than an extraterrestrial can determine the age of the earth by measuring the lifetime of say the longest living human being on earth. He will get a ridiculous value of 100 years!

So far astrophysics has helped us to have a general understanding of the evolution of the higher atomic number chemical elements in the cosmos as a dialectical process. Only a similar understanding of the genealogical nature of the large-scale distribution of the galaxies, their origin, morphology, evolution, dissolution, *etc.*, the energetics and the dynamics of the cosmos, the evolution of hydrogen, and other very low mass unit particles that are not created by the fusion reaction in the stars and so on; equipped with the consciousness of the laws of dialectical thought, will link us up with our understanding of the terrestrial nature. A cosmology built on the fantastic notions of a first cause, mathematics and a creator; a cosmology perplexed to explain how there can be galaxies even at 17 billion light years away, far beyond the limit of the universe it predicts, or how the ratio of heavy elements like iron in the intergalactic space and in the quasars (which are supposed to contain only primordial hydrogen and helium) can be as high as that in the galaxies, *etc.*; a cosmology that gives spurious explanation for the formation and the structure of the galaxies, or invokes mysterious monsters like dark mass/energy, black holes *etc* to explain the energetics/dynamics of the cosmos; and above all a cosmology that titters to the brink with nervousness on the outcome of a single experiment; will finally be done away with.

Ambartsumian's revolutionary insight marks a point of departure (a nodal point) for further progress in cosmology. It is a dialectical opposite of the paradigm of causality and of the single act of creation of all things that natural science fostered so long—a paradigm that has decidedly been proven wrong in the case of terrestrial nature. A paradigm that has progressively been built on professional careerism, conformity to the tradition and a sense that going against the trend is equivalent to cutting at the very branch of the tree you are sitting on. Only practicing visionary astrophysicists like Ambartsumian and a Darwin of the cosmos can rescue us and can help us out of the black hole that official cosmology has led us into. The rest of us can only hope and wait in anticipation like the poet Rabindranath Tagore:

> *Oh the fresh, the raw, the breaking light!*
> *Save the half-dead with thy fatal strike!*

* This article first appeared in the Internet based journal Apeiron, vol. 12, no.2, 2005, pp. 256-271.

THE COSMIC GAMMA-RAY HALO-NEW IMPERATIVE FOR A DIALECTICAL PERSPECTIVE OF THE UNIVERSE *

Abstract From the galaxies to man, natural science has so far sketched in the main outline an evolutionary picture of nature that is mediated by chance and necessity. Cosmic gamma-ray halo and the other observed cosmological phenomena must therefore, be viewed as the manifestation of the same evolutionary process of nature.

Keywords: Gamma ray halo, "Big Bang" theory, dialectical perspective.

Introduction

A halo of gamma-ray originating in and surrounding the Milky Way galaxy extends into outer space. Dave Dixon and co-workers reported this finding in a meeting of the High Energy Astrophysics Division of the American Astronomical Society, held on Nov. 4, 1997, at Estes Park, Colorado. This discovery (1) along with few other cosmological phenomena such as the cosmic microwave background radiation, the accelerated expansion of the universe, quasars etc., comprise the observational base, which any theory of cosmology has to contend with.

The existence of a gamma-ray halo surrounding Milky Way galaxy (and possibly other galaxies as well) is a unique astronomical observation in the sense that it cannot be attributed to a past event in the universe such as the "Big Bang" or to a compact source in the galaxy. This phenomenon must then be an ongoing *process* occurring at present throughout the *whol*e Milky Way galaxy.

There are at least three distinct cosmic gamma ray sources a) the known (and generally accounted for) sources at the galactic centres,

52

b) the diffuse gamma ray source occurring at large distances from the galactic core, and c) random gamma ray bursts (GRBs).

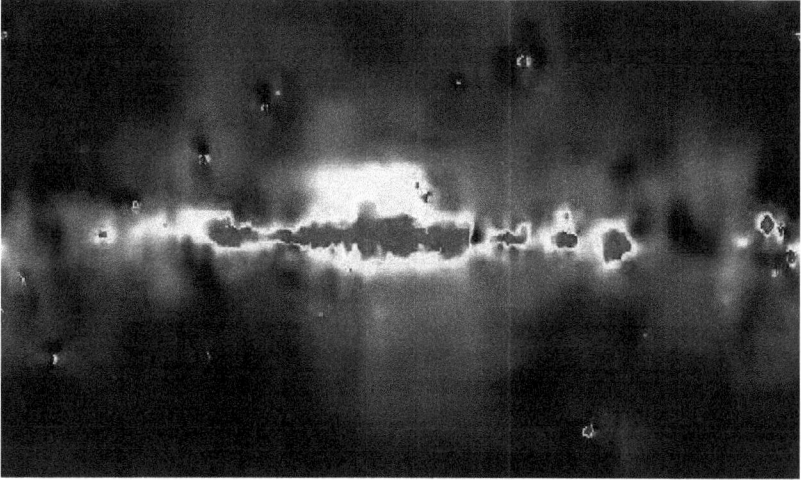

The Milky Way's Gamma-Ray Halo (Dixon et al.)

In this false colour all-sky gamma ray image (by Dixon et al) of the Milky Way shown above, the brown and green regions indicate brighter, known sources of gamma rays. The galactic centre and plane are clearly outlined, as are some distant galaxies seen near the top and bottom of the picture. The dim, blue regions above and below the plane correspond to the unexpected gamma-ray halo. According to Dixon et al. these gamma ray halo provides the first evidence that some sort of high-energy process is occurring at large distances from the galactic core. These processes probably also occur at the galactic core in addition to the known more dominant processes for generating gamma rays.

Gamma ray bursts are short (~10 ms to ~100s) random flashes of gamma rays at cosmological distances. Most of their energy is emitted around 1 MeV, with only a small fraction in the X-ray band. According to the "fireball" model, GRBs are generated by very high-energy electrons and positrons produced in an initial explosion

[2]. The most probable value of the peak energy of (redshift-corrected) GRB emission in the rest frame is reported to be very close to the electron-positron pair rest mass of 1022 keV [3].

Explanations of the diffuse extragalactic halo based on the currently accepted "Big Bang" paradigm, that the gamma-ray halo supposedly have its origin in the mysterious "dark matter" or in rapidly spinning neutron stars or is due to the interaction of cosmic rays with lower energy photons from the galaxy (an inverse Compton effect) are only after-thoughts.

According to the "Big Bang" theory, the universe was created about 15 billion years ago through a cataclysmic explosion. Space-time, matter and all that exists in the universe emerged from a primordial entity about the size of a proton through this single event and it is expanding ever since. The discovery by Edwin Hubble that the galaxies are receding from one another, the observation of a residual microwave background radiation and the relative abundance of low mass chemical elements such as hydrogen, helium etc. are in accordance with the "Big Bang" theory. This is presently the only viable theory of the universe and is purported to follow from the General Theory of Relativity propounded by Albert Einstein in 1915.

Philosophical Implications:

The "Big Bang" hypothesis, as well as the General Theory of Relativity is based on a line of thought which G.W.F.Hegel termed *Metaphysical* or the view of *understanding,* as opposed to his *dialectics* or the view of *reason. Understanding* views the world as a complex of ready-made *things* "created at one stroke" and "perfect in themselves". Nature in this view is essentially deterministic and a product of cause and effect, necessarily requiring a "first cause" or a creator; it has extension in space and only undergoes cyclic changes (if there are changes at all) but has no successive stages of *development* in time. *Dialectics* or *reason* on the contrary insists that the world must be comprehended as a complex of *processes* in which things go through an uninterrupted

54

change of "coming into being and passing out of existence" brought about by an interplay of chance and necessity and that they develop in successive stages. This distinction first came into sharp focus in natural science in the difference of views of Isaac Newton and Immanuel Kant about the cosmos. Through his theory of "cosmic evolution" Kant pointed to the dialectical development of the cosmos, that the celestial bodies and everything they contain *evolved* in time and that nature has a history not only of coexistence in space but also of development and succession in time. This was in sharp contrast with the previously held Newtonian view of the "perpetual" and "harmonious" motion of the heavenly bodies after an initial divine impulse. Our present understanding of the evolution of the solar system and the evidence for the evolution of the galaxies as a "bottom up" or a clustering rather than a "top down" or a fragmentation process [4] vindicates the dialectical views of Kant.

Darwin's theory of evolution and more recent developments in quantum mechanics, astrophysics, geology, palaeontology, biology and all other terrestrial sciences, point to a nature that is in eternal motion of evolution, development, change, brought about through a conflict of the opposites residing together in its elements, and that although at narrow particularity cause and effect has a role, at all levels of generality nature is governed by blind chance events, but with the iron necessity that is inherent in chance.

Dialectical View of the Universe:

In the light of the present development of quantum mechanics, it is probably possible to extend the intuitive ideas of early dialectical thinkers and those of Hegel, and to look at the universe as a *process*, as something that "comes into being and passes out of existence". If one naively assumes the universe to be an infinite void in which matter spontaneously "comes into being" in the form of some fundamental particles from "nothing" and similarly vanishes into "nothing" in the literary sense of the first Hegelian triad of "being-nothing-becoming" then there must always be some finite matter in the universe, because Uncertainty Principle forbids a perfect vacuum. If this spontaneous appearance and disappearance of matter is an eternal and everyday phenomena of innumerable "free lunches" instead of the quantum mechanically much less feasible one-time mighty "big bang" "free lunch", then the dialectical view of the universe becomes realistic and all of these fantastic mathematics and awe inspiring cosmology of "Big Bang" genesis becomes unnecessary. If the appearance and disappearance of matter is facilitated, so to speak catalyzed by the presence of existing matter as the graininess of the universe suggests and as quantum electrodynamics indicates the increasing concentration of virtual particles close to an atomic nucleus, and if such matter particles collects (like water molecules in the cloud) under the attractive force of gravity to form nebulae, galaxies, stars and so on then we can appreciate the anticipating poetic wonderment of Lucretius : "How is it that the sky feeds the stars!"[5].

Further, gravity cannot be only an attractive force, but according to the dialectical law of the *unity of the opposites*, must also possess a repulsive nature [6]. If the repulsive force of gravity at long distance overwhelms the relatively short-range attractive force, then the general dispersion of matter as observed by Hubble and the acceleration of this dispersion as observed recently, can be explained without invoking a primordial push from a "Big Bang".

The speculation about the creation and disappearance of matter as elementary particle, if true, must involve both matter and antimatter. If in one tiny region of the infinite void, matter gets pre-eminence over antimatter by purely chance events but with a necessity that chance entails, then the development of an island universe like the one in which we live in to be composed of only matter is perfectly feasible, because any anti-matter that forms is continuously eliminated through reaction with existing matter by the well known annihilation process, producing gamma-rays, – a sort of "natural selection" as is the case in the biological systems.

The observed gamma-ray halo may possibly be attributed to such an annihilation process going on in the Milky Way galaxy and its surrounding dust, because the creation of new matter and antimatter is more likely to be facilitated at centres where patches of matter already exist in the universe. The intensity of such gamma rays originating from a galaxy may, therefore, be proportional to its mass. The recent report by Caleb Scharf and Reshmi Mukherjee [7] that galaxy clusters are the main source of gamma ray in the visible universe is in conformity with such a possibility. The symmetrical nature of the gamma-ray halo with respect to the mass concentration in the Milky Way Galaxy, and other nearby galaxies observed by Dixon et al. also indicates the proportionality of gamma-ray intensity with the mass in this galaxy.

The origin of the high-energy radiation (X-rays, gamma rays etc.) at the galactic core with its wide emission band is usually attributed (among other things) to Bremstrahlung caused due to the super acceleration of galactic mass falling into the massive black holes. However, in such a case one would probably expect to observe relatively more compact sources due to the presence of a limited number of such black holes in the galaxy. But as the above image shows, the more intense gamma radiation from the galactic core and its immediate surrounding (the brown and green regions respectively) are also diffuse and symmetrical along the whole stretch of the galactic plane like the distant blue halo. The diffuseness, the proportionality of the gamma ray intensity with the increasing galactic mass concentration and its broad emission band is consistent with the above hypothesis that appearance – disappearance and the subsequent particle–antiparticle annihilation of evolving elementary matter is proportional to the existing mass distribution in the galaxy.

Another strong support of the above hypothesis comes from the discoveries of Halton Arp and others on the high-energy radiation from the center of the Local Supercluster [8] and on the quasars [9]. The strongest high-energy radiation (X-rays, gamma rays and ultra high-energy cosmic rays) in the sky coming from the direction of the Local Supercluster, was shown by Arp et al. [8] to originate from the centrally located Virgo Cluster with its associated Active Galactic Nuclei (AGNs) such as 3C274(M87), 3C273, 3C279 and Makarian 421. The creation-ejection of new mass from the active galactic core in the form of Planck-energy particles and quanta has been suggested as a possible explanation for this radiation source.

If the hypothesis of spontaneous appearance and disappearance of matter as elementary particles is correct, then one would not expect to observe any dramatic appearance or disappearances of ponderable galactic mass. It would be rather like a living animal whose cells die and new ones form as an ongoing process. Disappearance of galactic matter would not be observed because it would leave no trace, however newly appearing matter may, chance-accumulate for some time to a critical mass, when it can manifest itself in a catastrophic way as in the quasars reported by Halton Arp [9]. More over, as seems to be the case, one is more likely to observe the origin of the quasars at the galactic core, because according to this hypothesis new matter is more likely to form where already there is concentration of mass. It may be that the origin of the high-energy sources, the GRBs and the quarsars are (in different scale) manifestations of the same the same basic on going process occurring at the galactic core and it's surrounding dust.

The energy created by the matter-antimatter annihilation process and through other processes must decay during its lifetime to form a background radiation or so to speak, some "zero-point" energy in the universe also mandated by quantum mechanics. This "zero-point" energy can represent the cosmic microwave background radiation, which is touted as the incontrovertible proof of the 'big bang' theory.

Conclusion:

If such a simple picture of the universe is correct then it will be similar to the development of galaxies, solar systems and life or other processes in terrestrial nature, in all of which science so far has found to follow dialectical laws of development. This naïve but essentially correct way of looking at the world by the early dialectical thinkers, (and which can be perceived by any individual with some reflection), is being more and more reinforced with the development of natural science. The discovery of the gamma-ray halo in the Milky Way galaxy is a new and a significant affirmation of the dialectical view of the world.

References

[1] D.D. Dixon, D.H. Hartmann, E.D. Kolaczyk, J. Samimi, R. Diehl, G Kanbach, H. Mayer-Hasselwander and A.W. Strong, "Evidence for a Galactic gamma-ray halo", *NewA* 3 (No. 7, 1988) 539-561.

[2] G.J. Fishman, "Gamma-Ray Burst: Light from Darkness", *Nature*, 419 (2002) 259-261.

[3] I.G. Mitrofanov et al., "Comparison of Redshift – known Gamma-Ray Bursts with the Main Groups of Bright BATSE Events", *Ap. J.* 584 (February, 2003) 904-910.

[4] J. Silk, *The Big Bang*, W.H. Freeman & Company. (1989) 169 pp.

[5] T. Dickinson, *Quoted in The Universe Beyond*, 3rd. ed. Firefly Books Ltd. (1999) 69 pp.

[6] M. Edwards Ed., *Pushing Gravity*, Apeiron, Montreal. (April 2002).

[7] C. Scharf and R. Mukherjee, *Ap. J.* in press, *Scientific American*, (16 August 2002)

[8] H.C. Arp, J.V. Narlikar, and H.-D. Radecke, "High Energy Radiation from the Center of the Local Supercluster", *Astroparticle Physics*, 6 (1997) 387-394.

[9] H. Arp, *Seeing Red - Red shifts, Cosmology and Academic Science,* Apeiron, Montreal. (1998).

---

*The article originally appeared in the Internet based journal. *Apeiron*, 10, No. 2, April 2003) 165.

Gravity – An Intrinsic Property of Matter! - A Qualitative Graviton – Orbital-Band Theory

Abstract: *All profound theories of nature, life, and society have some philosophical underpinning; the theories of gravity are no exceptions. The theories of gravitation of Isaac Newton and Louis Le Sage were based on mechanical materialism and British empiricism. Albert Einstein developed his geometrical theory of gravity based on idealist Neo-Berkeleyan "positivism" of Ernst Mach. But none of these theories provide, among other things, any tangible intuition into the development of discrete, quantized and the shell like structure of matter from the subatomic to the cosmic, that modern physics, astrophysics and astronomy are revealing in increasing details. A dialectical (and quantum mechanical) approach to gravity based on a concept of quantized graviton-orbitals provides an explanation for the cellular structure, the discordant redshifts, and other observed cosmological phenomena.*

Introduction:

Matter in motion (movement, change, development, evolution etc) and its dialectical opposite inertia as expressed by among other forces, the universal force of gravity are the attributes of all existence from the microcosm to the macrocosm.

But a coherent contemplation by humanity of the very large and the very small could only arise with the development of nature-philosophy which attempted to understand the phenomena and things in nature in terms of some rational causes inherent in nature, rather than a dependence on superstition, myth, mystery or the super-natural.

But nature-philosophy could resolve itself in the two possible opposing views based on the primacy of the stasis or the dynamis; the ethereal or the material, spirit or nature, mind or matter, idealism or materialism and developed through the interplay of the two helical strands of dialectical opposites. The one strand - formal logic based on causality or the view of *understanding* sees everything in the world as isolated, immutable, *things* or *finished products*, (created at a stroke by one all pervading supreme being), "perfect in itself", that only have extension in space but no *development* in time. *Motion* in nature according to this view is a mystery; it arises only through an impulse from an alien agency or God. It regards the opposites as mutually exclusive and absolutely cut off from each other. The Aristotelian law of *identity, contradiction* and the *excluded middle* is the canon of its procedure. The other strand - (dialectical logic) brought to its most developed form by Hegel (1770-1831 A.D.) views the world as a *process* and things in it as unstable *unity of the opposites* that *come into being and pass out of existence* and undergo change and development in discrete stages, brought about by the conflict of the *opposites* inherent in their own *units* and mediated by chance and necessity.

These two strands of philosophy have their roots in the almost simultaneously proposed and the dialectically opposed philosophies of Parmenides (515 – 450 B.C.) and that of Heraclites (544 – 483 B.C.) respectively; and necessarily leads to contradictory, conflicting or exactly opposite notions on all questions of ontology and epistemology, including those on natural science as well as on gravity.

In modern natural science, causality (or the view of *understanding*) may be associated with, absoluteness, continuity, determinism, "good old common sense" of sharply divided either/or, yes/no, positive/negative, good/bad, cause/effect etc., categories of formal logic, and the field continuum of classical mechanics and mathematics; while dialectics may be identified with probability, uncertainty and discrete leaps, evolution, changes and developments in nature etc. most sharply evident in the quantum phenomena and life processes. For the view of *understanding* an *effect* irreversibly, invariably and deterministically follow a *cause*. This seems to be true for very

62

simple, short term, gross, unconnected and isolated instances and from the point of view of classical mechanics and "good old commonsense" of everyday life in macroscopic scale Causality served natural science well in its early stage of development of classical mechanics, when things had to be studied individually, in isolation and in their simple dynamics, separated from their natural surrounding. But problem arises when the situations get progressively more complicated in the case of the microcosm of physico-chemical and biological sciences where things have to be studied in an assembly of enormous complexity and in their inter connectedness with the totality of the world, where *cause* and *effect* inter-penetrate each other in a reversible way and interchange their role; what is a *cause* now and here becomes an *effect* then and there and vice versa. In the case of biological processes and the micro-world of quantum mechanics casual relations break down completely and can only be understood from the point of view of dialectics which posits that the opposites reside together in dynamic unity and contradiction, quality and quantity inter-converts into each other and motion, change & development occurs through *the negation of the negation*.

The fundamental difference between these two worldviews gets its concrete expression in natural science in the concept of *matter* and *motion* and these have important implication for any theory of gravity. While Newton recognized matter as *real*, for Einstein matter is a particular representation of an all-pervading space-time continuum of reality ("Being" of Parmenides?). The particle in this view can only appear as a limited region in space in which the field strength or energy density of the continuum is particularly high. Motion in the view of both Newton and Einstein could only arise from an *impulse* from without - from God. For dialectics (and quantum mechanics) on the contrary, matter and motion are the fundamental elements and the primary conditions of all physical reality; *matter and motion* are inseparable – *there can be no matter without motion or no motion without matter.*

63

Starting from Democritus (~ 420 B.C.) and Leucippus (~ 440 B.C.), materialism conceived all material objects as temporary and dynamic structures that exist in empty space and are composed of *discrete* particles or *atoms*. Democritus is known to have famously said, "Nothing exists except atoms and empty space; everything else is opinion". The properties of matter were attributed to the inherent characteristics and the arrangement of these atoms. Natural science thrived on this basis and found renewed confirmation of this idea with the development of quantum mechanics. Galileo (1564 – 1642 A.D.) attributed gravity as an inherent property of matter like mass, volume etc. However, the far-fetched extension of the limitedly valid concepts of causality, field continuum etc. in natural science and its dependence on either/or dichotomy of causality and *understanding* was bound one day to bring its idealist view of nature to the forefront. This was realized with the field concept of space-time continuum proposed by Albert Einstein in 1915. In this view, matter, has no independent and discrete identity as assumed by atomic theory but represents a concentrated standing wave of the space-time field continuum, which causes the "particle" effect at the wave center. The space-time like continuum is supposed to be the medium (new ether) through which the effect of gravitation and electro-magnetic radiation propagate.

In the case of the cosmos, the regularity and the perpetual motion of the nearby heavenly bodies were noted from antiquity and even expressed in precise mathematical terms; but a coherent and a realistic understanding of this motion and the "force" that kept these bodies in their orbits could only come about with the development of classical mechanics and (mechanical) materialism starting from late fifteenth century Europe. In 1683 Isaac Newton summarized the revolutionary discoveries of Copernicus, Galileo, Kepler and others to give a reasonably quantitative description of this motion and the "force" in the form of the laws of gravitation. But a proper *mechanism* as required by causality and classical mechanics for "action at a distance" and some observations, notably, the advance of the perihelion of the planet Mercury, the equivalence of kinetic and gravitational mass etc. were at odd with Newton's theory of gravity.

Albert Einstein's theory of General Relativity (GR) proclaimed in 1915 provided a relatively better quantitative description of gravity in terms of a fundamentally different and a radical concept of the geometry of space and time. Einstein explained away "action at a distance" by denying the concept of "force" itself. According to Einstein, gravity is the natural effect of the warping of the space-time fabric through which the celestial bodies move. The gravitational effect is due to the presence of massive bodies like the sun that creates a potential well in the space-time fabric into which the other smaller bodies like the planets get trapped and follow their natural course of motion without any force holding them in their orbit! But neither Newton's nor Einstein's theory provides any easy intuition for an explanation of the quantized, discrete and shell-like structure in nature or a *mechanism* for the "force" of gravity as understood in the conventional terms.

Although a simple mechanical theory of gravity (pushing gravity), based on the kinetic energy of an (assumed) all pervading particulate ether or graviton medium in conformity with the mechanical-materialistic spirit of the time was proposed by George Louis Le Sage of Geneva in mid 18th century, GR became the most successful and almost unassailable theory of gravity. But what started as a mundane and humble effort to overcome some of the limitations of Newton's theory of gravity and to explain the mystery of "action at a distance", has over time become a potent scientific tool in reinforcing the crumbling ontological and the epistemological narrative of causality, theology and idealism. The "Big Bang" creation of the universe extrapolated from GR provides a vindication of theological Genesis and elevates this theory to the status of an absolute truth. Therefore, the task of modern natural science, like that of theology is to interpret and to reveal the workings of this absolute truth in the details of nature – a task which modern official physics, astrophysics and astronomy are very enthusiastic to fulfill.

But proud materialism in general and natural science in particular had to pay a very high price to the mathematical idealism of GR and continue to pay heavy premiums by accepting supra-material objects like the black holes, dark matter/energy etc. Not withstanding the intoxication of natural science with GR, the question remains: if space-time, black holes, dark matter/energy etc. affect or are affected by matter, do they also come under the concept of *matter* at all or are these Kantian "things-in-themselves" that we will never know? Modern official science is preoccupied in finding an ever better proof of GR with more and more tenuous experimentations - a practice seems to be induced by a lingering doubt about its validity. The successes claimed by the theory's supporters consist of its ability to retrospectively fit observations by using adjustable parameters and by invoking mystical objects & processes (like superluminal inflation) as the need arise. From this perspective, the geometrical model of GR may be viewed as the more sophisticated modern version of Ptolemy's (323 – 283 B.C.) epicycles that very successfully served astronomy of the ancient Greeks and medieval theology, until Nicolas Copernicus (1473 – 15430). What is more ominous, GR has become an ideologically driven necessity and an exquisitely beautiful piece of mathematical art hung around the neck of natural science that attracts all kinds of cobweb spinning mathematical mystics with their unbridled imagination. As J.D. Bernal (1) noted, "It is just in those sphere of science where the least knowledge exists that the strongest attempts are made to use science to bolster up ancient superstitions"

For much of the past century and till now, GR provided the main paradigm for the official narrative of the cosmos. But in spite of the manipulated and guided direction, the results obtained so far by astronomy and astrophysics point to an intuition (2, 3), which is the exact opposite of that provided by GR. It is seen that the same dialectical laws that govern terrestrial nature, asserts themselves in the cosmos as well. That, instead of condensing from an initial and universal diffuse state, the galaxies evolve and unfold through the ejection and dissipation of matter from within the existing galaxies, due to the conflict of the opposites within themselves. That, galaxies extend in family like groups, clusters, super clusters etc. That, the discrete, quantized and cellular nature of the structure of matter in the microcosm (as established by quantum mechanics) persists in the organization of matter all the way to the macrocosm of the galaxies and their clusters as well (4). The discovery of the quasars and results obtained by radio astronomy, high-energy astrophysics during the past few decades point to a cosmos where colossal turbulence and also regularity, order and chaos reside in dialectical unity of the opposites. The cellular structure of the cosmos, the chemical composition of the intergalactic media and the quasars; and the quantization in the physical distribution & in the redshifts of celestial objects (5) are greatly at odds with the notion of their origin in a "first impulse", or the redshift as a measure of their cosmological distance. The idealist trammels of a grand cosmic design of majestic beauty and serenity etc. envisioned by GR are nowhere to be found.

All profound scientific theories of the past found their rational validity in human practice, in technology and industry. With all the lavish attention focused on GR it has provided very little tangible use and for all practical purpose, Newton's theory of gravity continues to provide the essential basis for the explorations and the studies of the cosmos. However brilliantly successful as a geometrical theory of gravity that GR is, the fact remains that it is rooted in a naïve epistemology of causality, conditioned by empirical experience in everyday life and at the same time it undermines the materialist base on which natural science founded it self. As our experience in chemical, biological sciences and in quantum mechanics clearly demonstrates; the "good old commonsense" of causality, provide no ready intuitive tools for an understanding of the micro world and it also casts doubt on field continuum that was the hallmark of classical field theories including GR.

This doubt was expressed by none other than Einstein himself in a letter to his friend Besso: "I consider it quite possible that physics cannot be based on the field concept, i.e., continuous structure. In that case, nothing remains of my entire castle in the air, gravitation theory included, (and of) the rest of modern physics" [6]. Whatever the real nature of gravity may turn up to be, this sober prophecy of Albert Einstein is more likely to come true in the course of time as the crisis in official theoretical natural science gets more and more aggravated. This doubt in Einstein reflects his sober realization: i) of the profound potency and the revolutionary nature of the quantum phenomena and its devastating effect on classical materialism and Newtonian mechanics on the one hand and ii) the recognition of the limitation of his attempt to minimize this revolutionary impact in natural science by co-opting mathematical idealism and field continuum in the cosmological sphere; on the other.

In this sense, Einstein is to natural science what Immanuel Kant was to philosophy. Both made reactionary and opportunistic attempt to save the idealist/rationalist notion (of certainty, eternity, universality, a priority, indubitability etc.) of human knowledge when faced with the fatal onslaught and challenge from the inexorable new radical developments in empiricism, natural science, materialism, and also of dialectical materialism in the case of Einstein. Mathematics, particularly geometry, played a central role in this task for both Kant and Einstein; as it did for Plato, the originator of idealist philosophy.

Mathematics As the Basis of idealism in GR and in Natural Science:

The history of the development of mathematics, like all other sciences, shows their intimate link with the practical activities of man; his need for hunting, build shelter, agriculture, technology, to measure land, time, keeping track of the seasons, economic

68

exchange, wars etc. The oldest known mathematical tablets date back from 2400 B.C. But like all other branches of knowledge, (e.g., religion, the state, capital etc,) and a certain stage of its development, the mathematical laws abstracted from the real world become alienated as if coming from outside and to which the world has to conform.

But from the early Greeks up until now, there has always been a sharp disconnect between the idealist philosophy of mathematics and its actual practice at any particular time. The idealist view of mathematics of Pythagorous (~580 – 483 B.C.) and Plato (429 – 347 B.C.) was reinforced by the rationalists (Decartes, Spinoza,, Leibnitz, Kant etc.), who maintained that rationality (formal logic) is an innate faculty of human mind, by which incontrovertible truth and knowledge that is eternal could be perceived a priori, independent of empirical experience. They cited mathematics especially geometry as the preeminent example for this fact; for starting from self-evident truths and proceeding by rigorous rules, geometry leads to certain, objective, eternal, indubitable and necessary truths about the world, without any logical contradiction and these truths are independent of sense perception or empirical experience - as demonstrated by Euclid's (~ 300 B.C.) geometry. But the reality of the matter as we know too well now is that the "self-evident truth" and presuppositions of idealism and geometry only have a finite lifetime. The self-evident truth of the sun and the planets going around a fixed earth ruled cosmology only for few centuries. Euclid's geometry, which was considered as the supreme example of the exact &, eternal properties of the universe and which even formed the rock solid foundation of the philosophy of Immanuel Kant, was found by nineteenth century to be only one of many forms of possible geometry.

Even the materialists and empiricists held that all knowledge except mathematics come from sense perception and observation. But they never explained how mathematical knowledge comes about. Since Kant, they followed his fuzzy, ambiguous, and subjective view of mathematics but generally adopted the idealist view when pushed hard enough. With GR (at least) Einstein was very honest, "Our experience hitherto justifies us in believing that nature is the realization of the simplest conceivable mathematical ideas. I am convinced that we can discover by means of purely mathematical constructions the concepts and the laws connecting them with each other, which furnish the key to the understanding of natural phenomena. ... In a certain sense, therefore, I hold it true that pure thought can grasp reality, as the ancients dreamed" (7). The criticisms of his empiricist predecessors like Copernicus, Kepler. Galileo was even blunt, "The natural philosophers of those days were on the contrary most of them possessed with the idea that fundamental concepts and postulates of physics were not in the logical sense free inventions of the human mind but could be deduced from experience by *abstraction* – that is to say by logical means. A clear recognition of the erroneousness of this notion really came with the general theory of relativity," [8]

In natural science for example, there is a gulf of difference between *pure* mathematics, whose program is the *exact* deduction of consequences from logically independent postulates, and *applied* mathematics of approximation needed to make tentative conclusions. Natural science uses approximate data or empirical relations, which it then tries to fit in various ways to *analytic functions* of *pure* mathematics for a more general and greater insight into the problem at hand. The analytic functions, i.e., those whose Taylor's series converge in the neighborhood of a given point have *precise, smooth* (no sharp breaks) and other desirable mathematical properties, which helps to make more generalized conclusions about an object or phenomenon. But the results are valid only in the *narrow* range of the observed values for the argument. Another convenient property of *analytic functions* is that, such functions are known for all the values of their argument when their values in any small range of their argument values are known. Thus the idealist proposition that the laws of nature involve *analytic functions* leads to a complete mechanistic determination of the world based on their experimentally determined value at a narrow range only.

So, by using circular logic, one can try to find some (mathematical) *analytic function* with extra adjustable parameter into which some experimentally determined values and observed quantitative relations in a thing or a process could be fitted to make profound, unlimited and far-fetched inferences. It would probably be not too much of an exaggeration to point out such a situation in the case of GR, where the field equations, parameters, constants (such as gravitational constant G, the velocity of light c etc.) derived from observations in the terrestrial and solar environment are declared to be universally valid for all time and over the whole universe. But the quantum phenomenon is a sad reminder that questions the validity of such an enterprise. Because, great many of the difficulties in applying classical mechanics to quantum phenomena lies in this kind of naïve and over-simplistic application of idealized mathematics to the real world. As dialectics rightly asserts, every truth has limits; when extended beyond this limit it turns into its opposite or to an absurdity.

Another important issue of cosmology that has a direct bearing on the nature of gravity and which was never resolved by the philosophy of *understanding* in general and natural science, in particular is the question of infinity, and the contradiction of the finite and the infinite (9). Hegel always treated with contempt the "gentlemen who subtilize over this contradiction". In GR, Einstein tried to resolve this contradiction by simply denying infinity. He arbitrarily assumed that the universe must be finite. Because his geometrical approach to gravity becomes helpless and his Machean-philosophy based cosmology collapses in an infinite universe. If Mach's principle is followed, then an infinite universe means that the inertia and the mass of atoms etc. also become infinite. To keep the world as we see it now (inertia, mass, etc.); all Mach based cosmologies must have the universe started at a finite past and also must have a finite extension.

It is ironical that the two basic pillars of GR, namely causality and the geometrical method would have been questioned by the two most influential idealist philosophers of modern time; i.e., David Hume (1711 – 1776 A.D.) and G.W.F. Hegel (1770 – 1831 A.D.) respectively. Hume convincingly argued that causality is a defective method for philosophical inquiry, because it leads to the mystery of a *first cause* (the *effect* of a *cause* that is unknown); also, causality lacks *necessity* and *universality*. Hegel for his part rejected the geometrical methods as tools of philosophical cognition, because "they have *presuppositions*; their style of cognition is that of *understanding*, proceeding under the canon of formal identity".

Also, Hegel's dialectical approach would deny the validity of the definitive, absolute, all-inclusive, invariable etc. epistemological claims of GR about the ontological nature of the universe. For dialectics, the contradiction of epistemology & ontology, like the contradiction of the finite & the infinite (9) resolves itself through infinite leaps and discrete stages, in the never-ending extension of space, time and history, without ever terminating to a final stage or an absolute truth. This is true for the epistemology of any fleeting/transient/impermanent consciousness that *comes into being and passes out of existence* in the universe in general and for the present one on the planet earth in particular.

As is already evident by now, Einstein's "finite universe" is only limited by the power and the range of the telescope we can build and already galaxies can be seen at 17 billion light years away; far beyond the limit predicted by the Big Bang theory! And his assertion of no preferential frame of reference for the universe is now undone by the ultimate reference frame of all time the cosmic microwave background radiation

All the theories of gravity proposed so far including GR are based on the paradigm of causality and the *view of understanding* of official science' even though the more recent developments in natural science (particularly, biology, chemistry and quantum mechanics) collectively continue to undermine their validity and on the contrary point to the dialectical perspective. Immanuel Kant for the first time proposed a dialectical view of the cosmos through his "nebular hypothesis", asserting that the heavenly bodies evolve and come into being in course of time. But natural science never followed through his profound lead, even though most recent discoveries in astrophysics vindicate Kant's vision (2,3). Of most fundamental importance is the fact that any epistemology based on the view of *understanding* in general and the field continuum of GR in particular, is incapable of comprehending or accommodating a discontinuous, quantized structure of nature that modern physics, quantum mechanics and astrophysics is bringing into view in ever increasing details. What is more; both Le Sage's theory and GR alienated gravity from being a property of matter, as something external to it, that conditions matter and in turn is conditioned by it. It seems that any further understanding of gravity can only come through a *negation* of this alienation. Only a dialectical and quantum mechanical approach can accomplish this task.

A Qualitative Quantized-Orbital Theory of Gravity:

A qualitative theory of gravitation based on a model of chemical bonding is presented here in. It is assumed that gravitons (of near zero rest mass, charge 0 and spin 2 as predicted by Quantum Theory of Gravity) form an integral part of the elementary particles in nature. The gravitons as bosons are their own virtual particles and like the photons, are also their own antiparticles. It is assumed that the gravitons exist as a cloud (or a standing wave) around elementary particles in quantized orbitals in loop-like band superstructure (like electrons in an atom) and can extend over long ranges in the empty space around the particle

73

center. This speculative idea was inspired by the recent report [10] about the possibility of the existence of "General relativistic boson stars" and a related highly speculative idea that dark matter could be in the form of a single giant boson star in which galaxies are embedded.

It may be possible to describe these graviton orbitals by some Schrödinger type wave equations, however a formal quantitative mathematical description may be impossible due to the problems of renormalization and the occurrence of infinities. A strictly quantitative quantum mechanical description of atoms and molecule is possible only in the case of very few simple systems. A qualitative narrative of the structure and bonding is enough to understand most of chemistry and biochemistry. In the same way, it is hoped that a qualitative and empirical approach may help to develop an understanding of the celestial bodies in particular and the universe in general. It needs to be emphasized that gravitation is a cumulative attractive force that has no known repulsive counterpart like the electromagnetic force that is operative in chemical bonding. The nature of graviton bonding like the boson stars can probably be studied by using similar set of stellar structure equations to those used for more conventional stars. Particularly, the equations would probably be similar to those describing neutron stars and result in similar properties.

Being bosons, the gravitons can exist in large but finite numbers in the same orbital of an elementary particle, as a loop-like laser beam. The graviton band is stabilized through attractive forces with the particle center as well as with other concentric graviton bands around the same particle center, mediated by the exchange of virtual gravitons. When two elementary particles or atoms come close together, part of their gravitons in the superstructure delocalize over the two centers (just the same way as in a molecular orbital) and this leads to a gravitational attractive or exchange force between the particles; the same way as chemical, electromagnetic, electro-weak and strong nuclear forces operate.

74

This attractive force is directly proportional to the density of the gravitons and hence the masses of the two particles and is inversely proportional to the square of distance between the particle centers as expressed by Newton's law of gravity. In the absence of any other forces, these two particles will have the tendency to fall towards each other. The delocalization continues (like metallic bonding) over large number of particle centers and over patches of matter either solid, liquid or gas and can extend over galaxies and clusters of galaxies. Gravity is a cumulative force and is unscreened by any intervening mass. Gravitational force is always attractive even in the super-cluster range. At very large e.g. extra super-cluster range, this exchange force (like nuclear force) becomes inoperative or may even act as a repulsive force. The recent report [10] about the possibility of the existence of "General relativistic boson stars" is in conformity with the localization of gravitons and the possible existence of graviton bands or orbitals.

To operate over galactic range, the gravitons as quantum particles, must posses much faster than light velocities as implied in the experimental testing of Bell's inequality theorem and as proposed by Tom Van Flandern [11]. Gravitons can move along the orbital bands like electrons that move over long distance through conducting media. Over inter-galactic and cluster ranges gravitational force may manifested itself as an exchange force like the other three forces of nature, where the gravitons acts as their own virtual particles and have very large but finite ranges. At large distance from the center of mass and hence at lower potential energy levels, the graviton orbitals may have directional characteristics like the p, d, f etc. orbitals of electrons.

Patches of matter ranging from the subatomic to the super-clusters of galaxies are therefore, bathed in a non-homogeneous dynamic sea of gravitons, where the gravitons impart a cohesive/attractive force between particle centers that function in addition to the other forces of nature and it is the only dominant force in the long range. The graviton clouds may exist in quantized and concentric contours around such mass centers as

75

fundamental particles, planets, stars, galaxies and clusters of galaxies etc, that extend as a halo over a few orders of magnitude beyond the radii of these centers. Matter points around these massive centers are therefore arranged along the graviton orbitals in shell like structures as we observe in the satellite, planetary, galactic and galactic-cluster structures in the visible universe.

The strength of the gravitational force is proportional to the density of the graviton cloud and is the strongest at the core of the galaxies and their clusters. This force (probably at non uniform scale) may be operative from the subatomic to the cosmic range and may have qualitatively and quantitatively different characteristics at very high or very low graviton cloud concentrations.

Any mass-object or electromagnetic radiation (photons) moving with a relative velocity through the graviton clouds will feel a deceleration (viscosity drag) through the "Jaakkola" or other similar effects that will be proportional to the cloud density, the mass and the velocity of the moving object or photons. Toivo Jaakkola [12], proposed an electro-gravitational coupling (EGC) interaction to explain the cosmic redshift as opposed to the expanding cosmological effect. As will be discussed later, the concept of graviton orbital may have important implications regarding the observed anomaly of red-shifts of the galaxies and quasars and the validity of Hubble's Law ascribing redshift of galaxies to the cosmic expansion as proposed by the Big Bang theory.

This simple and qualitative idea in conjunction with the recent postulates by this author [2] & [3] about the spontaneous appearance and disappearance of matter/antimatter (probably in the form of hydrogen and some other minor low mass atoms) and the propagation of galaxies through ejection/dissipation of matter from the existing galactic core propelled by the large scale and catastrophic annihilation reaction of chance accumulated patches of matter and antimatter, or other still unknown processes of ejection, provide a rational basis for many

76

of the observational and empirical phenomena in the universe, which is infinite in space and time in the Galilean sense. Some of the observed phenomena in the short or planetary ranges, mid or stellar ranges, large or galactic ranges and very large or cluster and super-cluster ranges are discussed below:

A. Effects at Short (Terrestrial) Range:

I. A very small but finite additional attractive force due the graviton bonding may be operative in the binding of the baryons and leptons in an atom, in addition to the nuclear and electro-weak forces. This may also partly account for the quantum electro-dynamic forces, which is observed as the "Lamb Shift" and other effects at the atomic scale.

II. The Van der Waals and Casimir forces and the formation of plasma crystals observed in laboratory experiments can be explained at least in part on the basis of graviton- orbital bonding. This type of short-range graviton orbital formation may have major implications in the initial formation of HI and HII clouds from newly created matter/antimatter in the galaxies, leading to the formation of nebula, star clusters and the eventual evolution of the galaxies. At a certain region and in the absence of any intrinsic organizing force, these newly created HI atoms that are randomly scattered over vast distances within the galaxy may band along the operating graviton bands of the galaxy and are grouped into clouds through long range graviton bonding among the atoms. These atomic clouds eventually lead to molecular clouds where additional chemical and other electro-magnetic forces become operative. Further concentration and chemical transformation of these clouds is effected through their interaction with the existing clouds of dust/atom plasma and the catalytic processes of the plasma crystals, which leads to the formation of nebula and eventually to the formation of stars where nuclear fusion processes come into operation. In this scenario all the forces of nature play their part in the evolution of the galaxies where the gravitational force play the decisive role.

III. The equivalence of kinetic and gravitational mass

Near the surface of a planet, like earth and in the absence of resistance from the atmosphere, the kinetic motion of a body relative to the planet will feel a viscosity drag proportional to the mass and the velocity of the body and the density of graviton cloud through which it is moving. Since for a finite range near the surface of the earth the graviton-cloud density may be assumed to be constant, the mass of a body will face the same drag whether it feels acceleration due to gravity or is propelled by an applied force. This will explain the equivalence of the kinetic and the gravitational mass; which was explained by invoking warped space-time in GR of Einstein This is also the reason why two bodies of different mass will fall at the same rate as in the famous experiment of Galileo.

B. Medium (Stellar) Range Effects

I. Planetary and Stellar Systems:

In the stellar and planetary nebula range quantized graviton orbital bands may in addition to other factors play an important role in the organization of the stars and planets in their orbits as a local phenomenon in the overall graviton cloud structure of the galaxy containing the nebulae. For the evolution of a planetary system, the size and mass of the mother nebulae, its chemical composition, dynamics (rotation etc.), graviton cloud structure etc. are major factors in the composition and the distribution of the planets and their satellites in their orbits. The quantized nature of the graviton cloud around a massive body like the sun may be an important factor in the distribution of the planets (and their satellites) around the star. As is discussed below, it is possible that a similar quantized or cellular structure may manifest itself in the organization of matter in all celestial structures like galaxies, and clusters of galaxies etc. A spectacular example of such a cellular structure even in the dust tail of Halley's comet is shown below:

Original plate taken with the UK Schmidt Telescope, David Malin, Nature, 320, 577 (1986)

II, The Advance of the Perihelion of the Planet Mercury::

The motion of Mercury in its orbit around the sun will be retarded (gravitational drag) in proportion to its mass, speed and the graviton density along its path. This effect will be the highest at the perihelion where the speed of Mercury and the graviton density is highest because of its proximity to the sun. This will cause a certain perturbation of the motion of Mercury and lead to the precision of the perihelion. This effect will be much smaller for other planets, which are at larger distances from the sun and revolved around it at lower speeds.

III. Time Delay of Signals (Shapiro Effect), Pioneer Anomaly, Gravitational Red Shifts:

All these effect have been observed experimentally. Although the Shapiro Effect and Gravitational Red Shift have been explained on the basis of GR and even modified Newtonian dynamics, the Pioneer Anomaly remains a mystery. Pioneer 10 and 11 were sent on missions to Jupiter and Jupiter/Saturn respectively. The calculated position of the Pioneers did not agree with measurements based on timing the return of the radio signal being sent back from the spacecraft. These consistently showed that both spacecraft were slowing down more than they should be, by thousands of kilometers If the Pioneer anomaly is a gravitational effect due to some long-range modifications of the known laws of gravity, it does not affect the orbital motions of the major natural bodies in the same way Hence a gravitational explanation would need to violate the equivalence principle of GR, which states that all objects are affected the same way by gravity. All these effects may be explained on the basis of the slowing down ("viscosity drag") of moving bodies and/or electromagnetic radiation in the dense graviton cloud of the sun and in the surrounding graviton orbitals in the extended halo.

IV. The Bending of Star Light:

The celebrated experiment on bending of star light by the gravity of the sun performed by Aurther Eddington in 1919 was purported to provide an incontrovertible proof of GR and warped space-time. But this may be a simple case of refraction of the star light in the graviton cloud of the sun due to a combination of Jaakkola, Wolf, Brillouin, Dynamic Multiple Scattering etc. type effects discussed later in the section on redshifts.

V The Dust Rings Around Supernova 1987A

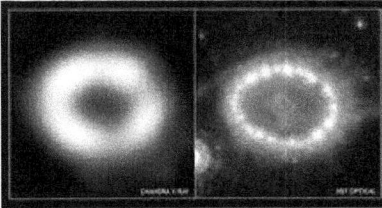

Supernova 1987A in X-ray and visible light.

Credit: X-ray: NASA/CXC/U.Colorado/S.Zhekov et al.; Optical: NASA/STScI/CfA/P.Challis

Supernova 1987A exploded in the Large Magellanic Cloud. Two sheets of dust near the Supernova deflected some light from the Supernova. The sheets of dust were observed as two concentric rings and were seen long after the star faded away, because the scattered light cover a longer path to reach the earth. The dust rings are about 470 and 1,300 light years away from the supernova center. Supernova 1987A belonged to a cluster of stars. It is possible to speculate that these gas rings existed along the graviton orbitals of this star cluster and were made visible through the scattering of the intense supernova radiation by these dust shells. No explanation for this ring structure based on GR or any other hypothesis is available.

C. Large (Galaxies and Groups) Scale Effects:

1. Dark Matter

No direct evidence for the existence of dark matter or its nature is available. Its presence is inferred from the motions of celestial objects. The orbital motion of the peripheral objects in an assembly (stars in a galaxy, galaxies in a cluster etc.) is too high for gravity of the visible mass of the assembly to retain them in place. Some additional matter (dark matter/energy) amounting to more than 90% of the total mass of the assembly is necessary to stop its disintegration. In addition, Gravitational lensing of distant galaxies is purported to show that there are far more matter than one observes as visible mass. The dark matter is also required in the big bang model to enable gravity to amplify the small fluctuations in the primordial atom (in the Big Bang scenario) enough to form the large-scale structures that we see in the universe today. Dark matter is supposed to be the dominant source of gravitational forces in the Universe.

The graviton orbital model eliminates the necessity of the mystical dark matter halo in the galaxies and their clusters. Graviton orbital like molecular orbital in a chemical systems provides the necessary glue and cohesive force to keep the patches of matter in galaxies and clusters in place and allow for the observed high rotation velocities of the objects at the periphery. In addition, the Newtonian gravitational constant (G) at any point in space may be a function of matter and graviton concentration at that point. At very high matter /graviton concentration such as at the core of galaxies or cluster of galaxies G may attain vastly different value through some resonance or synergistic effect. G measured in the terrestrial environment and extrapolated within the solar system is taken to be an universal constant all through the universe in both Newtonian and General Relativity theories. Moreover, if one accepts the view of Ambartsumian and the dialectical view proposed by this author that dispersion/ejection is the fundamental process in the evolution of the universe, (and not conglomeration of diffuse matter through the formation of virialized and dynamically relaxed structures); then the need of dark matter to keep galaxies from disintegration does not arise.

II The Structure of the Galaxies

The distribution of matter within the galaxies may have cellular or quantized structure because of the existence of graviton orbitals. This is obvious enough in spiral galaxies, where the spiral structure is stabilized over billions of years, even under a strong pull of the outer arms towards the center of the galaxy. It is possible that more compact galaxies such as E or SO types may also have similar but less obvious cellular structures. Deep photographic plates of elliptical galaxies do indeed reveal faint shells that extend out to 2/3 times further beyond the more luminous structures. NGC 3923 for example show extensive network of circular arc extending over 150,000 light years from the galactic center as shown below.

The shells of NGC 3923

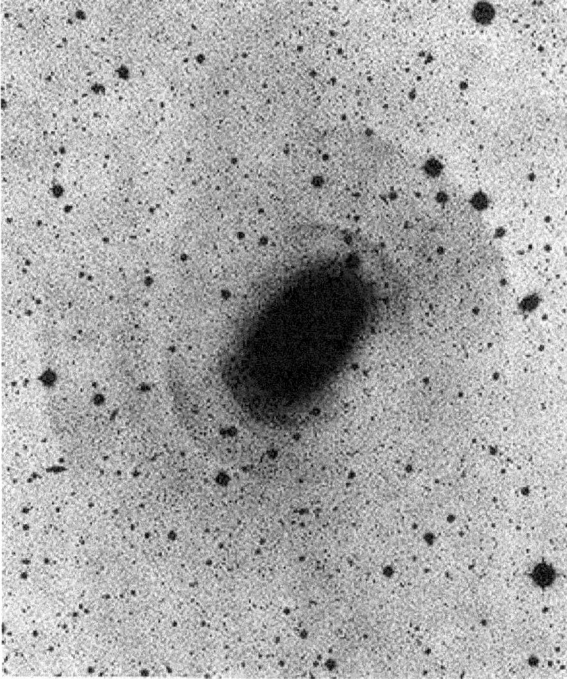

Photo by David Malin & David Carter on Anglo-Australian Telescope

This type of shell like structure seems to be a general phenomenon for elliptical galaxies. In a Catalogue of Elliptical Galaxies with Shells, David Malin and David Carter [13] lists 150 new examples and illustrates some unusual specimens, such as NGC 1344, NGC 3923, and NGC 1549-53, mostly discovered using photographic amplification on UK Schmidt plates.

More specific and detailed information on the galactic rings appeared recently in a catalogue, published in the *Monthly Notices of the Royal Astronomical Society*, that includes 113 such rings in 107 galaxies (14). Six are dust rings in elliptical galaxies, while the rest (the majority) are star-forming rings in disc galaxies. The nuclear rings are ring-shaped, star-forming configurations located around galactic nuclei, as shown in the example below:

The largest atlas of nuclear galactic rings has been unveiled. (Credit: NASA, ESA, D. Maoz, G. F. Benedict et al.)

The possibility that two nearby galaxies can interact through the overlap of their outer graviton orbitals is shown in the following photograph of galaxies NGC 1549 and NGC 1553. This picture shows striking similarities with the distribution of electron clouds in a diatomic molecule!

84

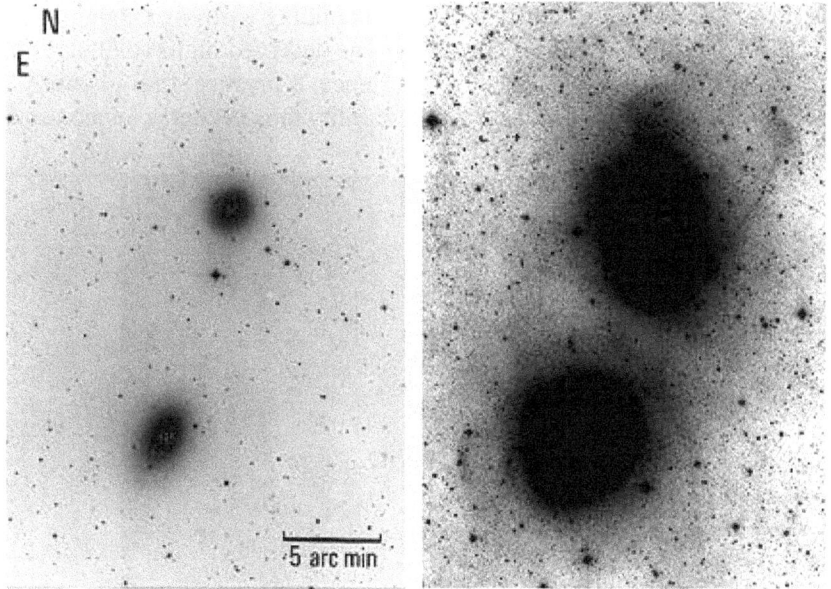

These galaxies were listed in the Malin-Carter catalogue of ellipticals with shells.

D. Very Large (Extra-galactic) Range Effects:

1. Clusters and Super Clusters Formation

It is conceivable that galaxies in groups, clusters, superclusters etc. interact with each other through graviton-orbital bonding (as it seems evident from the above picture of NGC 1549 and NGC 1553) to form dynamic but shell-like structures with the satellite clusters arranged in a hierarchy along the graviton orbitals of the central dominant clusters. This is dramatically demonstrated in the recent photograph of Abell 2218 (picture bellow), a rich galaxy cluster composed of thousands of individual galaxies. It sits about 2.1 billion light-years from the Earth (redshift 0.17) in the northern constellation of Draco. The ring like structure of this galaxy cluster have been explained in GR model as due to gravitational lensing of background galaxies by the dominant foreground cluster producing multiple images. It is possible that even in large clusters, graviton bonding is strong enough to stabilize these structures with the observed high radial motion of the objects at the periphery. Hence, a mystical "dark matter" to account for this anomalous stability thus becomes unnecessary.

Abell 2218 *(Credit: NASA, ESA, and Johan Richard (Caltech, USA))*

Below is a composite optical and X-ray image of a Super cluster pair

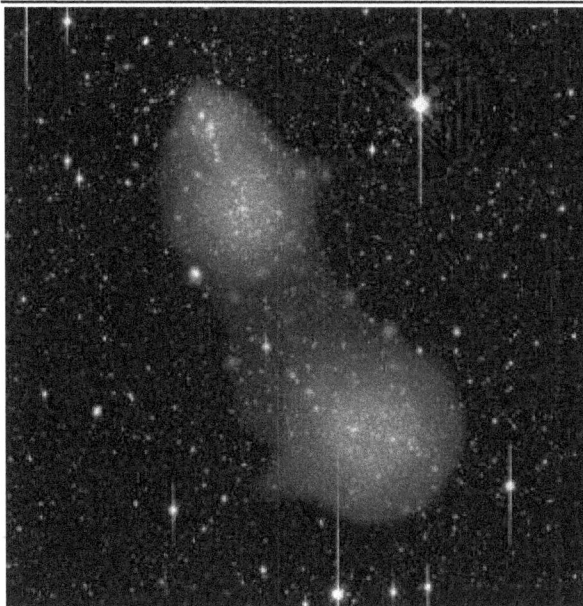

Composite optical and X-ray image of galaxy clusters Abell 222 and Abell 223. The cluster pair is connected by a filament permeated by hot X-ray emitting gas. (Credit: ESA/ XMM-Newton/ EPIC/ ESO (J. Dietrich)/ SRON (N. Werner)/ MPE (A. Finoguenov))

At the range of super clusters and beyond, galaxies appear to be distributed in filaments, strings and in thin sheets that surround large seemingly empty bubbles or voids up to 300 million light years in diameter. It is possible that at very large distances from the central mass the graviton orbitals possess directional characteristics like the p, d & f etc. atomic orbitals. Alternatively, the gravitational interaction at large (extragalactic range) may take the form of an exchange force involving virtual gravitons and the bonding between clusters acquire directional characteristics due to very large distances involved, giving rise to string or filament like structures. Out side the range of exchange force, gravitational interaction becomes inoperative and may even act as a repulsive force representing the so-called "dark energy"

E. Quasars and Anomalous Red shifts:

Quasars represent the biggest embarrassment of the big bang theory and other similar cosmological theories based on mathematical idealism and the general theory of relativity. Since it was first suggested by Halton (Chip) Arp (15) few decades ago that high red shift quasars are ejects from nearby (low red shift) active galaxies, mountain of rapidly accumulating observational evidence including quasar – galaxy associations, close pair of quasars, their alignments and groupings, red shift periodicities and quantized redshifts effect etc. [5 &15] is making breeches in the high walls of closely guarded big bang paradigm that obstinately refuges to accept [16] the ejection theory. In addition, the lack of explanation for the high content of metals (iron etc.) in the quasars, inter-galactic medium & in the "infant universe"; such enormous red shifts of the quasars based on the Hubble relation that impose impossible non-relativistic recess velocity (more than the speed of light); measurable proper motion and enormous energy requirements for quasar luminosity at such great distance etc. is making the big bang paradigm more and more untenable. Renewed debate about the quasar redshift as a measure of distance is raging even in the "big bang camp" itself [17].

The recent report [18] of the discovery of a high redshifted quasar very close to the nucleus of NGC 7319 and the other observational evidence accumulated so far support the contention of Ambartsumian, Arp, Hoyle, Narlikar, Burbidge and others [5, 19] that the quasars must be looked at as ejects of explosions from nearby large active spiral galactic centers; that their red shifts are intrinsic and not due to cosmological effect caused by the expansion of space-time and recess velocity, as big bang theory contends. A satisfactory resolution of the quasar phenomena may have profound implications for modern cosmology.

The fact that quasars (as opposed to the radio galaxies) appear to be single jet systems, and quasars always appear as red shifted was explained in the 1960s, by Strittmatter (cited in Ref. 5), who suggested that quasars selectively radiate in the backward direction as it is ejected from the galactic centers and an observer therefore, only sees those quasars that are receding from that location and are necessarily red shifted. Kembhavi and Narlikar [5] gave an explanation for this effect in terms of the ram pressure of the intergalactic medium on the light originating from the quasars. According to them, the emission from the quasars is confined to a backward cone; the ram pressure from the intergalactic medium blocks the forward emission, but it does not affect the backward jet. This explains why the quasars unlike radio galaxies are single jet systems and are always redshifted.

Early on in the quasar controversy, Hoyle and others [20] attempted to explain the redshifts of the quasars in terms of the effect of ejection velocity related relativistic aberration, intrinsic Doppler and gravitational effects due their very high ejection velocity. Although the view that the quasars are high velocity ejecta from active galactic centers explains some of the observed phenomena, an explanation for the very large redshifts and the quantization phenomena still remain elusive. The large red shift is explained on the basis of a novel "variable mass" theory first proposed by Hoyle and Narlikar [21] in the 1960s. More recent work by Narlikar and Das [22] describes the dynamics of ejection of quasars by galaxies in which the ejecta from the galactic centers consists of newly created low-mass matter (with large red shift) which eventually grow into normal matter to form galaxies with regular red shifts. No direct evidence for this hypothesis from astrophysical observation or from particle physics is yet available.

A possible satisfactory understanding of the quasar phenomena and the large red shift may be obtained from a consideration of the very high graviton-cloud density at the core of the galaxies and the electro-graviton coupling interaction proposed by Jaakkola [12]. It was proposed recently by this author [3], that quasars are chunks of nearby active galaxies expelled with enormous speed along the minor axis of the spiral galaxies powered by the large scale annihilation of matter with antimatter that chance-accumulate at the galactic core. In such a scenario, the very high redshifts of the quasars are intrinsic and caused by a combination of Doppler Effect due to the enormous recession velocity of the quasars, the gravitational and the large Jaakkola effect (viscosity drag) on the emitted light during ejection phase. In such a case the relativistic aberration, intrinsic Doppler and gravitational effects on the quasar red shift (as proposed by F. Hoyle et al.) will be further accentuated due to the Jaakkola effect of electro-graviton coupling of the backward emission from the quasar with the dense and extended but non-homogenous graviton cloud of the mother galaxy. This may explain the large red shifts of the quasars even though they are ejected from nearby galaxies. The quasars represent the first short-lived stage of the satellite galaxies that surround the large active galaxies. In time the quasars are slowed down due to the gravitational pull of the mother galaxy to form satellite galaxies with normal redshifts that are conveniently located along the dominant graviton bands of the mother galaxy.

In addition, the dense and non-homogenous graviton cloud in the halo of the ejecting galaxy may provide condition to enhance the Wolf Effect [23] and of correlation-induced spectral shifts. Recently Roy et. al. [24] proposed the Dynamic Multiple Scattering Theory (based on Wolf effect) that "provides mechanism for non-cosmological red shift of quasars and a simple interpretation of the discordant red shifts in galaxy-quasar associations". They used this theory to provide additional support for the specific galaxy-quasar association of the classic case of NGC 4319 and Markarian 205, investigated by H. Arp and J.W. Sulentic.

Further more, the non-homogenous graviton clouds in the galactic halo may induce Brillouin type scattering of the quasar emission thereby increasing the redshift. In classical physics, Brillouin scattering occurs when light interacts with density

90

variations as it passes through a media such as water and changes its path. From a quantum mechanical point of view, Brillouin scattering is explained by the interaction of light photons with the phonons (vibrational quanta) of the media. The non-homogenous graviton medium in the galactic halo may provide similar scattering mechanism of the quasar light.

In addition to the above factors, light from each celestial object is likely to have a unique path history as it passes through the dense graviton clouds from the extended halo of galaxies in its path and the Milky Way galaxy itself and also the light absorbing intergalactic media as proposed by Marmet [25]. Therefore, a distance related (tired light) redshift effect may also be in operation. A combination of measurable contribution to the red shift from all the above factors may be enough to explain the observational evidence of Arp and others that high red shifted quasars are ejects from nearby low red shifted galaxies with active galactic nuclei and invalidate Hubble's distance-redshift proportionality.

F. The Quantization of Redshifts and Cellular distribution of Celestial Objects:

The great "division of labor" in modern physics that quantum mechanics must deal only with the microcosm and GR with the macrocosm is so complete that it required a rare stroke of luck and the great insight of some relatively under-funded astrophysicists to show that the *continuum* of GR does not rule the cosmos at large. From astronomical observation and from a study of their redshifts it was discovered [4, 5, 15, 16] that the quasars (G. Burbidge, 1968), galaxies (W. Tifft, 1987), and the clusters of galaxies (J. Einasto, 1998) show quantized peaks in their physical distribution as well as in their redshift values.

The notion that matter should exist in discrete, quantized structures in the cosmic scale as it is proven to be the case at atomic scale; should come as no surprise from a dialectical point of view, because dialectics asserts that there is no (one-time)

91

leap in nature like the Big Bang, it consist entirely of infinite and discrete leaps. The graviton-orbital model explains in a natural way the quantized arrangement of the quasars, galaxies and the clusters of galaxies. If the quasars and satellite galaxies are ejecta from central and active galactic nuclei or are captured in the gravitational field of the central galaxy by chance encounter, then these satellites will preferentially stabilize along the graviton-orbital bands of the central dominant galaxy or cluster. If it is assumed that the ejection of matter/energy takes place along a narrow cone around the minor axis of spiral galaxies (3) and that a galaxy must reach an optimum size before it develops an active nuclei, then the position of the quantized graviton-orbital in the galactic halo, the escape velocity and the mass of the ejecta etc. may give a net effect such that its equilibrium distance from the mother galaxy will be constrained to a limited range. This will produce distribution peaks for their distance from the mother galaxy and also a quantization of their redshifts. The emission of the ejecta in the quasar phase will show higher redshift because of Jaakkola and other effects discussed above and eventually will stabilize itself preferentially along the graviton-orbital bands of the central galaxy with regular redshifts.

References:

[1] J.D. Bernal, "Dialectical Materialism and Modern Science", *Science and Society*, II, (No. 1, 1937).

[2] A. Malek, "The Cosmic Gamma-Ray Halo – New Imperative for a Dialectical Perspective of the Universe", *Apeiron*, 10, No. 2, April 2003) 165.

[3] A. Malek, "Ambartsumian, Arp and the Breeding Galaxies", *Apeiron*, 12 (No. 2, April 2005) 256.

[4] H.Arp., "Seeing Red: Redshifts, Cosmology and Academic Science", Apeiron, Montreal, (1998) 195.

[5] A.K. Kembhavi & J.V. Narlikar, *Quasars & Active Galactic Nuclei*, Cambridge University Press, (1999) 419

[6]   A. Pais, Subtle is the Lord ... " *The Science and the Life of Albert Einstein"*, Oxford University Press, (982) 465,

[7]   A. Einstein, "Essays in Science", Translated by Alan Harris from "Mein Weltbild, Quedro Verlag, Amsterdam, 1933), The Wisdom Library, N.Y., p48 – 49, (1934).

[8]   A. Einstein, ibid.,

[9]   A. Malek, "The Infinite - as a Hegelian Philosophical Category and Its Implications for Modern Theoretical Natural Science", Mukto-Mona, Feb. 05, (2009).

[10] F.E.Shunck and E.W.Mielke, "General Relativistic Boson Stars", *Class. Quantum Grav.*, 20 (2003) R301-R356.

[11] T.V. Flandern, "Gravity" in *Pushing Gravity*, M.R. Edwards Ed. Apeiron, Montreal, (2002) 93.

[12] T. Jaakkola, "Action-at-a-Distance and Local Action in Gravitation", *ibid,* 155 pp.

[13] D. Malin and D. Carter, "A Catalogue of Elliptical Galaxies with Shells", *Ap. J.*, 274, (1983) 534-540. *Nature*, 320 (1986) 577.

[14] S. Comerón, J. H. Knapen, J. E. Beckman, E. Laurikainen, H. Salo, I. Martínez-Valpuesta, R. J. Buta. **AINUR: Atlas of Images of NUclear Rings**.*Monthly Notices of the Royal Astronomical Society*, 2010; 402 (4): 2462 DOI: 10.1111/j.1365-2966.2009.16057.x

[15] Arp, H. *Catalogue of Discordant Redshift Associations,* Apeiron, Montreal, (2003).

[16] Hawkins, E et. al., *MNRAS,* 336 (2002) L13.

[17] Schilling, G. "New Results reawaken quasar Distance Dispute", *Science,* 298 (11 Oct. 2002) 345.

[18] Galianni, P. et.al., "The Discovery of a High Redshift X-Ray-Emitting QSO Very Close to the Nucleus of NGC 7319", *Ap.J.*, 620, (Feb. 10, 2005) 88.

[19] Arp, H.C. et.al, "The Extragalactic Universe, an Alternative View", *Nature*, 352, (1990) 807.

[20] Hoyle, F. and. Fowler, W.A "Gravitational Redshifts in Quasi-Stellar Objects", Nature, 213 (1967) 373.

[21] Hoyle, F. and Narlikar, J.V. "A Conformal Theory of Gravitation", Proc. Roy. Soc. Lond. A294 (1966) 138.

[22] Narlikar, J.V. and Das, P.K. "Anomalous Redshifts of QSOs", *AP.*, 80 (1982) 127.

[23] Wolf, E. "Noncosmological redshifts of Spectral Lines", *Nature,* 326, (1987) 363.

[24] Roy, S. et al., "Shift of Spectral Lines due to Dynamic Multiple Scattering and Screening Effect: Implication for Discordant Redshifts", *Astron. Astrophys,* 353, (2000)) 1134.

[25] Marmet, P. "The Cosmological Constant and the Redshifts of Quasars," *IEEE, Transactions on Plasma Science*, Vol. 20, (No. 6, Dec. 1992) 958.

APPENDICES

A. Comments in *Nature* News-blog, "In the Field":

1. Cosmic Ripples Net Physics Prize:

Nobel Prize for Big Bang Theory Raises Big Questions. Is Nobel Prize in Physical Sciences Losing Some of Its Nobleness?

The recent award in support of the Big bang theory seems to be a notable point of departure for Nobel awards in Natural Sciences. While Nobel awards for World Peace, Economics and even Literature can be construed to be based on subjective (ideological, politico-economical etc.) considerations of "Western" interest, awards for Natural sciences always remained beyond any trace of controversy.

Cynics, for example can point out that, during the past few decades at least, Nobel World Peace prize was awarded for ideological consideration against the "Evil Soviet Empire" [Andrei Sakharov – 1975, Lech Walesa – 1983, Mikhail S. Gorbachev – 1990] and for politico-economic interest in the case of the Middle East [Menachem Begin and Anwar el-Sadat -1978, Yasir Arafat, Shimon Peres and Yitzhak Rabin – 1994, Shirin Ebadi – 2004]. Even the aptly deserving World Peace award to Jimmy Carter – 2002 was in general perceived to be a reaction against recent American policies in the Middle East.

Nobel awards in Natural sciences on the contrary always merited almost universal consensus, especially from those working in the field. These awards were always characterized by strict adherence to objectivity, the importance of the discovery and an uncompromising demand for verifiable experimental results. Albert Einstein was awarded Nobel Prize for his work on photoelectric effect, but not for his theories of Special and General Relativity, even though these two theories dominated the practice of Natural science for the last 100 years. Another notable example is the idea of "Black Hole", even though vigorous claims about its existence are being made for a long time. The Theory of General Relativity (GR) still remains a field of vigorously active research and a major preoccupation of modern physics; astrophysics and cosmology; where increasingly tenuous attempts are being made to prove GR with even better accuracy – a passion apparently guided by a lingering doubt about its validity and the necessity to support related theories such as Big bang by proxy.

Both GR and the Big bang theory that has its scientific root in the former were dogged with controversy and scepticism from the moment they were proposed. It is a historical fact that Arthur Eddington over-enthusiastically claimed to have proven GR based on partial results of his experiments on the bending of star light by the gravity of the Sun. Calculations based on the complete data recorded by his team did not support the initial claim by Eddington. Although later experiments by other researchers verified the bending of star light by the Sun, this episode set a bad precedent which prompted Stephan Hawking to observe, "a case of knowing the result they wanted to get, not an uncommon occurrence in science".("A Brief History of Time", Bantam Books, p-32, (1988)). Einstein's initial insertion of a fudge factor (for a static universe) in his equation and later claim (after Hubble's discovery of an expanding universe) that it was the "biggest blunder of his life" was also unhelpful. All the sophisticated and difficult experiments (including COBE) done later on were perceived by many to be contrived and seemed to have been carried out in response to new requirements to keep the theories alive.

COBE based study of the microwave background radiation is the most sophisticated experiment ever done on this subject. This required expensive and elaborate logistical, technical, data processing etc. expertise and instruments that had to work at the very limit of their technical abilities. A study of this nature can only be carried out by a resourceful government supported agency like NASA. A repetition of any aspect of this study seems unlikely in the near future and remains outside the ability of any other group of scientists. Although GR commands total loyalty from most established and mainstream physicists, this is not the case with Big bang theory. Opposition to the officially accepted Big bang theory was voiced on empirical and epistemological grounds by a significant fraction of astrophysicists and cosmologists since 1950s and it is still growing. Ambartsumian, Arp, Alfven, Hoyle, Narlikar, Burbidge pair, Phipps are only few names among many who strongly contested the Big bang theory. Evidence for quasar-galaxy association, shell like structure of galaxies and their clusters, quantization in the physical distribution and in the redshifts of the galaxies, existence of heavy metals like iron in the intergalactic space and the quasars which are supposed to contain only primordial gas, the very high redshifts of the quasars that require much faster than light non-relativistic velocities, their unrealistic luminosity and proper motions at such enormous distances (in the Big bang scheme), observation of galaxies at 17 billion light years away – far beyond the limit predicted by the Big bang theory etc., are at odd with its paradigm.

Astronomer and astrophysicist Halton Arp, one of the most famous opponents of the Big bang theory has this to say in a personal communication in reference to the present Nobel Award, "The intergalactic medium has to have some temperature. Eddington calculated about 2.7 deg. in 1926. In the 1940's Max Born calculated 2.7 deg. on the basis of tired light. Gamow calculated 50 deg. before Pezias and Wilson measured 2.73 deg. But a Canadian astronomer, McKellar had already measured it from the excitation of the inter-stellar CN molecule. The ultimate irony is that it is a primary reference frame which violates Einstein's assumption about no preferred reference frames." - And on George Smoot's work on the small variation in the temperature of the microwave background radiation, "Eric

Lerner in his latest summary describes how, that "face of God" (a remark made by Smoot after his discovery – AM) is now falling apart. The predictions stepped down a number of times before any irregularities finally showed up at 10^-5."

The claims of the Big bang theory at best are being contested and at worst are uncertain. The Nobel Award of this nature seems to be an exception to the tradition. According to the Nobel citation, the Big bang "is the only scenario that predicts the kind of cosmic microwave background radiation measured by COBE". Even though Per Carlson, chairman of the Nobel physics committee said according to press reports that "they have not proven the Big bang theory, but they give it very strong support"; this award represents a very strong endorsement of the Big bang theory and very likely will further stifle healthy scientific debate on this issue. This endorsement adds to the one already given by the Vatican.

Posted by: Abdul Malek | December 21, 2006 11:58 PM
http://blogs.nature.com/news/blog/2006/10/cosmic_ripples_net_physics_pri.html

2. APS April 2008: Fermilab could rule out one type of Higgs:

The Hunt for the "God Particle"

Near religious status of the Theory of General Relativity (GR) and the reductionist paradigm of modern physics have effectively reduced it into an exclusive goal seeking enterprise; undermining the empirical basis, which was once the greatest strength of natural science. Near complete monopoly of funding and facilities for research; heavy reliance on political and governmental agencies have rendered natural science ever more vulnerable to politico-ideological manipulation on the profound ontological questions of reality at macrocosm and microcosm.

98

The hunt for Higgs boson like COBE is such a goal seeking effort that has profound implications for humanity. Natural science must be extremely careful in making any definitive claims on these fundamental ontological questions – questions that may forever be elusive to human epistemology.

This is particularly important in a politically charged atmosphere where strong subjective efforts are being made to bring back medieval absolutism, by striving to bring an "End of Science", "End of History" etc. based on some "absolute truths" of natural science itself.

Any positive claim (no matter how tenuous) on the discovery of Higg's boson; like the COBE's claim of the Big Bang creation of the universe is certain to bring Nobel, Templeton etc. awards and severely limit further debate on these fundamental issues. I have tried to draw attention to these questions in some recent articles (links below"):

http://blogs.nature.com/news/blog/2006/10/cosmic_ripples_net_physics_pri.html

Posted by: Abdul Malek | May 7, 2008 05:12 PM

B) Comments in the Guardian Newspaper:

1. The neutrino may prove the true revolutionary image of 2011

futurehuman's comment 22 November 2011 7:58PM

The new revolution (after the Copernican revolution) in natural science that was engendered by the revolution in philosophy by Hegel, Marx and Engels, the ideas of evolution, and by quantum mechanics will come to fruition with the undoing of Ensteinian mathematical idealism. In this sense the new results on neutrino speed and possible negative results on the existence of Higgs boson both associated with the Large Hadron Collider (LHC) may indeed turn out to the "greatest subversives" and the last straw for the already crisis ridden monopoly capitalism.

Newtonianism did not "die in 1919" as this author suggest. It is true that Einstein's theories were born in the revolutionary period of early 20th century, but contrary to general belief it was a reactionary effort to reinforce Newtonianism, against the new revolutionary developments cited above, to save the certainty of causality against the immediate threats of dialectical philosophy and quantum mechanics.

In this sense, Einstein is to natural science what Immanuel Kant was to philosophy. Both made reactionary and opportunistic attempt to save the idealist/rationalist notion (of causality, certainty, eternity, universality, a priority, indubitability etc.) of human knowledge when faced with the fatal onslaught and challenge from the inexorable new radical developments in empiricism, natural science, and also of dialectical materialism in the case of Einstein.

Einsteinism coincided with the transition of capitalism to its present fully developed monopoly stage, and played a similar ideological role for new monopoly capitalism, as that played by Martin Luther for revolutionary capitalism in early 16th century.

With the decline of the influence of theology, monopoly capitalism latched on to Einsteinian mathematical idealism for its ideological underpinning and for strengthening idealism and theology. High value experiments (starting from the dubious early experiment of Aurther Eddington) during the past few decades were carried out to "prove" the theories of relativity with ever more tenuous and sophisticated (as if there were doubts with the results of the previous ones) experiments. And unsurprisingly ALL of them confirmed the theory "with flying colors". Three Noble Prizes have been awarded so far, for "proving" the Big Bang theory – an extension of the theory of General Relativity (GR).

The LHC (initiated by the most passionate and, the greatest champions of monopoly capitalism - Ronald Reagan and Margaret Thatcher) was meant to seal the deal by finding the God Particle, before the "Final Theory", the "Theory of Everything", "The end of history" etc., and the crowning victory of monopoly capitalism are proclaimed.

But it now seems that the irony of dialectics and the results of the LHC may bring the exact opposite of the intended goal for monopoly capitalism.

www.ingramcontent.com/pod-product-compliance
Lightning Source LLC
Chambersburg PA
CBHW060419090426
42734CB00011B/2366